Forum for Interdisciplinary Mathematics

Forum for Interdisciplinary Mathematics is a Scopus-indexed book series. It publishes high-quality textbooks, monographs, contributed volumes and lecture notes in mathematics and interdisciplinary areas where mathematics plays a fundamental role, such as statistics, operations research, computer science, financial mathematics, industrial mathematics, and bio-mathematics. It reflects the increasing demand of researchers working at the interface between mathematics and other scientific disciplines.

More information about this series at http://www.springer.com/series/13386

G. Hariharan

Wavelet Solutions for Reaction–Diffusion Problems in Science and Engineering

 Springer

G. Hariharan
School of Arts Science
and Humanities (SASH)
SASTRA Deemed to be University
Thanjavur, Tamil Nadu, India

ISSN 2364-6748   ISSN 2364-6756   (electronic)
Forum for Interdisciplinary Mathematics
ISBN 978-981-32-9959-7   ISBN 978-981-32-9960-3   (eBook)
https://doi.org/10.1007/978-981-32-9960-3

This Springer imprint is published by the registered company Springer Nature Singapore Pte Ltd.
The registered company address is: 152 Beach Road, #21-01/04 Gateway East, Singapore 189721, Singapore

Dedicated
With Love and Regards to
My Parents and Wife

Preface

The primary intention of this book is to examine the efficiency of various wavelet methods when applied to multiple problems of nonlinear and fractional-order reaction–diffusion equations of substantial importance. Having an easy-to-fallow scientific insight and being sufficiently realistic for studying important design problems, reaction–diffusion models of enzyme kinetics play an important role in chemical kinetics theory. The characterizing equations of enzyme kinetics models are highly nonlinear reaction–diffusion equations which do not have analytical solutions. As existing methods can only handle a limited range of these equations, many computational methods have been developed in recent years, having either equal or better performance. In general, the qualitative behavior of the solutions may not always be fully exposed by semi-analytical method results. In order to achieve this goal, discrete wavelet transform is studied first, followed by their properties, convergence, and computational complexity for addressing a few issues of enzyme kinetics. Therefore, this book investigates theoretically a few steady- and unsteady- state reaction–diffusion problems arising in enzyme kinetics models.

Wavelet method is a recently developed tool in applied mathematics. Investigation of various wavelet methods for their capability of analyzing various dynamic phenomena has gained more attention in engineering research. Starting from offering good solutions for differential equations to capturing the nonlinearity in data distribution, wavelets are used as appropriate tools at various places to provide a decent mathematical model for scientific phenomena, usually modeled through linear or nonlinear differential equations. Review shows that the wavelet method is efficient and powerful in solving wide class of linear and nonlinear reaction–diffusion equations. This book also intends to provide great utility of wavelets to science and engineering problems which owe its origin to 1919.

Chapters maintain a balance between mathematical rigor and practical applications of wavelet theory, thereby, catering to students and researchers with particular needs, wanting to understand not only the reaction–diffusion problems but also wavelets theory in order to have a broader understanding. Operational matrices have been introduced to convert the given nonlinear and fractional differential equations into a system of nonlinear algebraic equations. Applications of Haar,

Legendre, and Chebyshev wavelet methods and wavelet-based hybrid methods in the field of nonlinear and fractional-order reaction–diffusion equations are also included for the first time. This book also includes innovative techniques for finding the approximate solutions of highly nonlinear boundary value problems. Wavelet-based methods have been used to combine the strength of both analytical and numerical methods to produce efficient hybrid techniques.

When compared to other numerical methods of solutions, discrete wavelet transforms (Haar, Legendre and Chebyshev) have some preferences such as mathematical efficiency, simplicity, and possibility to implement standard algorithms and high accuracy for a small number of grid points. Solutions based on the wavelet methods are usually simpler and faster than in case of other methods. For these reasons, wavelets have obtained greater popularity and the number of papers about discrete wavelets is rapidly increasing. For a reader, it is difficult to find his way among a large number of publications.

Therefore, a book like this, explaining the applications of the discrete wavelet transform in calculus, is extremely necessary. As different variants of the wavelet methods exist, it is not reasonable to handle and analyze all of them in detail. Therefore, we have decided to choose a method of solution, which is sufficiently universal and is applicable to solve all the problems by a unit approach. Other treatments will be referred to and discussed in the section-related papers added to each chapter. To demonstrate the efficiency and accuracy of the proposed method, a number of examples are solved. Mostly test problems, for which the exact solution or solution obtained by other methods is known, are considered.

The book is meant for researchers, teachers, and students of applied mathematics, physics, engineering, and related disciplines. To make the book accessible for a wider circle of readers, some mathematical finesse is left out.

Thanjavur, Tamil Nadu, India   Dr. G. Hariharan

Acknowledgements

I take this opportunity to express my sincere gratitude to Prof. R. Sethuraman, Chairman, SASTRA University, for his constant support, encouragement, and the excellent academic and research atmosphere provided. I wish to thank Dr. S. Vaidhyasubramaniam, Vice-Chancellor, SASTRA University, for his manifold plans to develop the research activities of hardworking researchers to set new goals in interdisciplinary areas. His sound managerial principles coupled with his future vision have been of great help to the mathematics research community. I express my sincere thanks to Dr. S. Swaminathan, Dean, Planning and Development, SASTRA University, for his constant support and encouragement.

It is not out of place to acknowledge the efforts of my Ph.D. scholars who worked with me on my DRDO-NRB and DST-SERB research projects. The research work related to this field has greatly inspired me to write this book.

I also express my sincere gratitude to Dr. K. Uma Maheswari, Dean, School of Arts, Sciences and Humanities (SASH), and Dr. R. Srikanth, Associate Dean/SASH, for their kind cooperation during the preparation of the book.

I have received considerable assistance from my colleagues in the Department of Mathematics, SASTRA University. Moreover, I am especially grateful to the team of Springer for cooperation in all aspects of the production of the book.

I thank my parents for their blessings. Last but not least, I thank my wife and my daughters for their patience and support.

I am looking forward to receiving comments and suggestions on this work from students, teachers, and researchers.

Contents

About the Author

G. Hariharan has been Senior Assistant Professor at the Department of Mathematics, SASTRA University, Thanjavur, Tamil Nadu, India, since 2003. He previously served as a lecturer at Adhiparasakthi Engineering College, Melmaruvathur, Tamil Nadu. Dr. Hariharan received his MSc and PhD degrees in Mathematics from Bharathidasan University, Trichy, and SASTRA University, in 1999 and 2010, respectively. He has over 20 years of teaching experience at the undergraduate and graduate levels at several educational institutions, as well as 16 years of research experience in applied mathematics. He is a life member of the Indian Society for Technical Education (ISTE), Ramanujan Mathematical Society (RMS), International Association of Engineers (IAENG), and the Indian Society of Structural Engineers (ISSE).

Dr. Hariharan has served as the principal investigator of projects for the DRDO-NRB (Naval Research Board) and Government of India, and has contributed research papers on several interdisciplinary topics such as wavelet methods, mathematical modelling, fractional calculus, enzyme kinetics, ship dynamics, and population dynamics. He has published over 85 peer-reviewed research papers on differential equations and their applications in various leading international journals, including: *Applied Mathematics and Computation*, *Electrochimica Acta*, *Ocean Engineering*, *Journal of Computational and Nonlinear Dynamics*, *MATCH-Communications in Mathematical and Computer Chemistry*, *Aerospace and Space Sciences*, and the *Arabian Journal for Science and Engineering*. In addition, Dr Hariharan serves on the editorial boards of several prominent journals, including: *Communications in Numerical Analysis*, *International Journal of Modern Mathematical Sciences*, *International Journal of Computer Applications*, and *International Journal of Bioinformatics*.

Nomenclature

b   Microbial death constant, $cm^3/(mg\ day)$
D_f Diffusion coefficient within the biofilm, cm^2/day
J   Substrate flux into the biofilm, $(mg\ cm^2)/day$
K   Michaelis–Menten constant, mg/cm^3
L_f Biofilm thickness, cm
q   Substrate consumption rate constant, day^{-1}
S   Dimensionless substrate in the biofilm
S_f Substrate concentration in the biofilm, mg/cm^3
S_L Dimensionless substrate concentration outside the biofilm
S_I Substrate concentration outside the biofilm, mg/cm^3
T   Time, days
x, y   Dimensionless coordinate, cm
X_t Concentration of physiologically active microorganisms, mg/cm^3

Chapter 1
Reaction–Diffusion (RD) Problems

1.1 Reaction–Diffusion Equations (RDEs)

Reaction–diffusion equations (RDEs) are nonlinear parabolic Partial Differential Equations (PDEs). RDE arises in many applications which include physical sciences, biological sciences, ecology, physiology, finance, to name a few. Reaction–diffusion systems are usually coupled systems (multiple numbers) of parabolic partial differential equations. In population dynamics, the reaction term models growth, and the diffusion term accounts for migration. A few reaction–diffusion (RD) models are models for pattern formation in morphogenesis, for predator–prey and other ecological systems, for conduction in nerves, for epidemics, for carbon monoxide poisoning, and for oscillating chemical reactions.

A simplest form of RDE:

$$u_t = \frac{\partial u}{\partial t} = D\frac{\partial^2 u}{\partial x^2} + f(u)$$

where $u = u(x, t)$ is the vector of dependent variables, $f(u)$ is a nonlinear vector-valued function of u (the reaction term), and D is the diffusion coefficient. The reaction term arises from any interaction between the components of u. The parameter u may be a vector of predator–prey interactions, competition, or symbiosis. The diffusion terms may represent molecular diffusion or some 'random' movement of individuals in a population.

A simplest form of reaction–diffusion–convection type is given by

$$\frac{\partial u}{\partial t} = u_t = f(u) + D\frac{\partial^2 u}{\partial x^2} + C\frac{\partial u}{\partial x},$$

where C is the convection coefficient.

The diffusion mechanism model is the movement of many individuals in an environment or media. The particles reside in a region, which we call Ω is open set

© Springer Nature Singapore Pte Ltd. 2019
G. Hariharan, *Wavelet Solutions for Reaction–Diffusion Problems in Science and Engineering*, Forum for Interdisciplinary Mathematics,
https://doi.org/10.1007/978-981-32-9960-3_1

of R^n (the nth-dimensional space with Cartesian coordinate system) with $n \geq 1$. The diffusion coefficient $D(x)$ is not a constant in general since the environment is usually heterogeneous. But when the region is approximately homogeneous, we can assume that $D(x) = D$, the above equation can be simplified to

$$\frac{\partial P}{\partial t} = D\Delta P + f(t, \ x, \ P),$$

where $\Delta P = \text{div}(\nabla P) = \sum_{i=1}^{n} \frac{\partial^2 P}{\partial x_i^2}$ is the Laplacian operator.

In order to develop reaction–diffusion models as dynamical systems, we need to define appropriate state spaces of functions and determine how the models act on them.

1.2   Importance of Reaction–Diffusion (RD) Problems

(i) **Chemical Engineering**: Theoretical models of steady- and unsteady-state reaction–diffusion problems have been developed to obtain the substrate and product concentrations for enzymes immobilized within particles. Reaction–diffusion models are characterized by carbon monoxide poisoning, nitrogen oxide removal, oscillating chemical reactions, pulse splitting and shedding, Rayleigh–Benard convection, and kinetics of methylene blue adsorption (film–pore diffusion model). A theoretical model based on the Michaelis–Menten enzymatic conversion of the substrate and the diffusion of the sub-strate was created. They also describe the steady-state oxygen diffusion in a spherical cell and equilibrium of isothermal gas sphere, flame propagation, autocatalytic chemical reactions, and neutron population in a nuclear response and branching.

(ii) **Biological and Medical Sciences**: A few important applications of reaction–diffusion equations include population dynamics models, gene propagation models, ecological invasions, a spread of epidemics, tumor growth, and wound healing, distribution of heat sources in a human head, transmission of pulses in nerves, and neurophysiology.

(iii) **Mechanical Engineering**: A simplified kinematical description of a rigidly rotating spiral induced in a general two-component reaction–diffusion medium is elaborated by application of a free-boundary approach. The potential energy generated by an external force as a result of a deformation is propagated among mass points by the principle of reaction and diffusion.

(iv) **Civil Engineering**: A theoretical model based on fundamental reaction–diffusion principles has been formulated to describe the process of concrete carbonation. It is a major time-limiting factor for the durability of reinforced concrete.

1.3   A Few Familiar Reaction–Diffusion Equations (RDEs)

1.3.1   Nonlinear Singular Boundary Value Problem (Lane–Emden Type) and Wavelets

Nonlinear singular boundary value problem (Lane–Emden type) is a significant model in the theory of stellar structure. It models many phenomena in mathematical physics and astrophysics. Most of the work in the stellar structure was initiated by Chandrasekhar [1]. It is a nonlinear differential equation which describes an equilibrium density distribution in the self-gravitating sphere of polytrophic isothermal gas and has a regular singularity at the origin. This model equation was first studied by the astrophysicist Lane [2] who considered the temperature variation of a spherical gas cloud under the mutual attraction of its molecules and subject to the laws of classical thermodynamics. The polytrophic theory of stars was studied by Davis [3]. It primarily deals with the issue of energy transport, through the transfer of material between levels of the star and modeling of clusters of galaxies. Mostly, problems with regard to the diffusion of heat perpendicular to the surfaces of parallel planes are represented by the heat equation. In particular for a polytropic equation of state, the Lane–Emden equation arises.

Due to the simplicity, the wavelets are very effective for solving ordinary differential and partial differential equations [4–9]. Therefore, the idea, to apply wavelet technique also for solving reaction–diffusion problem, arises. The wavelet methods with far less degrees of freedom and with smaller CPU time provide better solutions than classical ones [10–19]. The accuracy and effectiveness of the method are analyzed; the results obtained are compared with the results of other authors (using classical numerical techniques) and with the exact solution, evaluating the error.

1.4   Fractional Differential Equation (FDE)

Fractional calculus is a field of mathematical study that deals with investigations and applications of derivatives and integrals of noninteger orders. In recent years, fractional differential equations have been applied for efficient models in research areas as diverse as dynamical systems, control systems, mechanical systems, chaos, anomalous diffusive and subdiffusive systems, continuous time random walks, wave propagation, and so on.

1.5   Definitions of Fractional Derivatives and Integrals

(i) **Riemann–Liouville fractional operator**:

$$
{}_a^R D_t^\alpha f(x) = \begin{cases} \dfrac{\mathrm{d}^m f(t)}{\mathrm{d}x^m}, & \alpha = m \in N, \\[2ex] \dfrac{\mathrm{d}^m}{\mathrm{d}x^m} \dfrac{1}{\Gamma(m-\alpha)} \displaystyle\int\limits_0^x \dfrac{f(t)}{(x-t)^{\alpha-m+1}}\mathrm{d}t, & 0 \le m-1 < \alpha < m. \end{cases}
$$

(ii) **Caputo fractional operator**:

$$
{}_a^C D_t^\alpha f(x) = \begin{cases} \dfrac{\mathrm{d}^m f(t)}{\mathrm{d}x^m}, & \alpha = m \in N, \\[2ex] \dfrac{1}{\Gamma(m-\alpha)} \displaystyle\int\limits_0^x \dfrac{f^{(m)}(t)}{(x-t)^{\alpha-m+1}}\mathrm{d}t, & 0 \le m-1 < \alpha < m. \end{cases}
$$

Time-fractional reaction–diffusion equation is given by

$$
\frac{\partial^\alpha u}{\partial t^\alpha} = D\frac{\partial^2 u}{\partial x^2} + R(u).
$$

If $\alpha = 1$, it is a standard diffusion, if $0 < \alpha < 1$, it is known as anomalous subdiffusion, and if $1 < \alpha < 2$, it is known as anomalous superdiffusion. In anomalous subdiffusion, diffusive behavior is associated with highly damped particles in random potential, and in anomalous superdiffusion, diffusive behavior is associated with low damping with periodic potential.

1.6   Mathematical Tools to Solve Fractional and Nonlinear Reaction–Diffusion Equations

Many problems of the physical phenomena in science and engineering are governed by differential equations and integral equations. In order to understand the behavior of the aspects, it is necessary to find exact solutions for such problems. But, in general, it is very difficult to find such an accurate analytical solution due to complexity while modeling such equations. Many researchers are involved in developing various methods both numerically and semi-analytically for finding solutions for such difficult problems which require complex derivates and integrals. The ordinary differential equations form the basis for many mathematical models such as physical, chemical, and biological phenomena, and more recently, their use has also spread into economics, financial forecasting, image processing, and other fields. So, examination of such mathematical models often needs numerical approximation where an analytical solution does not exist. As there is no direct method to solve specific equations like Fisher's and FitzHugh–Nagumo (FN) equations in population dynamics, Lane–Emden-type equations arise in astrophysics, film–pore diffusion model in chemical engineering, etc. Hence, approximate methods can be used for finding solutions of this nature. In most cases, an approximate solution is represented by functional values at specific discrete points or grid points or mesh points. To get the better approximation of solutions with smaller error, several numerical methods have been developed in the literature to solve nonlinear differential equations. A few semi-analytical techniques are Adomian decomposition method (ADM), homotopy analysis method (HAM), homotopy perturbation method (HPM), variational iteration method (VIM), and differential transform method (DTM). The main idea behind the above techniques is to obtain an approximate solution for the differential equation by expressing the unknown function regarding infinite series of polynomial basis. One of the significant drawbacks of the polynomial basis approach is that it does not indicate up to what extent the unknown function can be expressed regarding polynomials. In order to overcome the difficulty in above methods, a recently invented tool in applied mathematics called wavelets has been used to solve such kind of difficult problems.

(a) Adomian Decomposition Method (ADM)

The Adomian decomposition method is an important semi-analytical method for solving ordinary and partial nonlinear differential equations. The main aim of this technique has been superseded by a more general theory of the homotopy analysis method.

Let the general functional equation be

$$y - N(y) = f. \tag{1.1}$$

where N is a nonlinear operator, f is a known function in which the solution y satisfying Eq. (1.1) is to be found. We assume that for every f, Eq. (1.1) has a unique solution. The Adomian technique consists of approximating the solution of Eq. (1.1) as an infinite series.

$$y = \sum_{n=0}^{\infty} y_n, \tag{1.2}$$

and decomposing nonlinear operator N as

$$N(y) = \sum_{n=0}^{\infty} A_n, \tag{1.3}$$

where A_n are Adomian polynomials $y_0, y_1, \ldots, y_n$ given by

$$A_n = \frac{1}{n!} \frac{d^n}{d\mu^n} \left[N\left(\sum_{i=0}^{\infty} \mu^i y_i \right) \right]_{\mu=0}, \quad n = 0, 1, 2, \ldots \quad (1.4)$$

Substituting (1.2) and (1.4) in (1.1) yields

$$\sum_{n=0}^{\infty} y_n - \sum_{n=0}^{\infty} A_n = f. \quad (1.5)$$

Thus, we can identify

$$\begin{aligned} y_0 &= f, \\ y_{n+1} &= A_n(y_0, y_1, \ldots, y_n), \quad n = 0, 1, 2, \ldots, n. \end{aligned} \quad (1.6)$$

We can then define the M-term approximation to the solution y by

$$\phi M[y] = \sum_{n=0}^{M} y_n,$$

with

$$\lim_{M \to \infty} \varphi M[y] = y.$$

(b) Variational Iteration Method (VIM)

The main property of the method is in its flexibility and ability to solve nonlinear equations accurately and conveniently. The variational iteration method was proposed by Ji-Huan He, which was successfully applied to autonomous ordinary and partial differential equations. The main advantages of this technique are the correction functions which can be constructed easily by the Lagrangian multipliers, which can be optimally determined by the calculus of variations. The application of restricted variations in correction functional makes it much easier to learn the multiplier. The initial approximation can be freely selected with possible unknown constants which can be identified via various methods. The results obtained by this method are valid not only for a small parameter but also for a very large parameter.

To clarify the basic ideas of He's VIM, we consider the following differential equation:

$$Lu + Nu = g(t),$$

where L is a linear operator, N a nonlinear operator, and $g(t)$ is an inhomogeneous term.

According to VIM, we can write down a correction functional as follows:

$$u_{n+1}(t) = u_n(t) + \int_0^t \lambda\left(Lu_n(\xi) + N\ddot{u}_n(\xi) - g(\xi)\right)d\xi,$$

where λ is a general Lagrangian multiplier which can be identified optimally via the variational theory.

The subscript n indicates the nth approximation, and $\ddot{u}_n$ is considered as a restricted variation, i.e., $\delta\ddot{u}_n = 0$.

(c) Homotopy Perturbation Transform Method (HPTM)

By the homotopy technique in topology, a homotopy is constructed with an embedding parameter $p \in [0, 1]$, which is considered as a 'small parameter.' To illustrate the basic idea of the method, we consider a general nonhomogeneous partial differential equation with initial conditions of the form

$$Du(x,\ t) + Ru(x,\ t) + Nu(x,\ t) = g(x,\ t), \tag{1.7}$$

$$u(x,\ 0) = h(x),\ u_t(x,\ 0) = f(x),$$

where D is the second-order linear differential operator $D = \frac{\partial^2}{\partial t^2}$, R is the linear differential operator acting on x of order less than D, N represents the general nonlinear differential operator, and $g(x, t)$ is the source term. Taking the Laplace transform denoted by L on both sides of Eq. (1.7):

$$\mathsf{L}(Du(x,t)) + \mathsf{L}(Ru(x,t)) + \mathsf{L}(Nu(x,t)) = \mathsf{L}(g(x,t)). \tag{1.8}$$

Using the differentiation property of Laplace transform, we have

$$\mathsf{L}(u(x,t)) = \frac{h(x)}{s} + \frac{f(x)}{s^2} - \frac{1}{s^2}\mathsf{L}(Ru(x,t)) + \frac{1}{s^2}\mathsf{L}(g(x,t)) - \frac{1}{s^2}\mathsf{L}(Nu(x,t)). \tag{1.9}$$

Operating with Laplace inverse on both sides of Eq. (1.9) gives

$$u(x,t) = G(x,t) - \mathsf{L}^{-1}\left(\frac{1}{s^2}\mathsf{L}(Ru(x,t+Nu(x,t)))\right), \tag{1.10}$$

where $G(x, t)$ represents the term arising from the source term and the prescribed initial conditions. Now we apply homotopy perturbation method (HPM)

$$u(x,t) = \sum_{n=0}^{\infty} p^n u_n(x,t), \tag{1.11}$$

and the nonlinear term can be decomposed as

$$Nu(x,t) = \sum_{n=0}^{\infty} p^n H_n(u), \tag{1.12}$$

for some, He's polynomials which are given by

$$H_n(u_0 \ldots u_n) = \frac{1}{n!} \frac{\partial^n}{\partial p^n} \left[N\left(\sum_{i=0}^{\infty} (p^i u_i) \right) \right]_{p=0}, \quad n = 0, 1, 2, 3, \ldots \tag{1.13}$$

Substituting Eqs. (1.11) and (1.12) in Eq. (1.10), we get

$$\sum_{n=0}^{\infty} p^n u_n(x,t) = G(x,t) - p\left\{ \mathsf{L}^{-1}\left[\frac{1}{s^2} \mathsf{L}\left[R \sum_{n=0}^{\infty} p^n u_n(x,t) + \sum_{n=0}^{\infty} p^n H_n(u) \right] \right] \right\},$$

which is the coupling of the Laplace transform and the homotopy perturbation method using He's polynomials. Comparing the coefficient of like powers of p, the following approximations are obtained.

$$p^0 : u_0(x,t) = G(x,t),$$

$$p^1 : u_1(x,t) = \mathsf{L}^{-1}\left[-\frac{1}{s^2} \mathsf{L}[Ru_0(x,t) + H_0(u)], \right],$$

$$p^2 : u_2(x,t) = \mathsf{L}^{-1}\left[-\frac{1}{s^2} \mathsf{L}[Ru_1(x,t) + H_1(u)] \right],$$

$$p^3 : u_3(x,t) = \mathsf{L}^{-1}\left[-\frac{1}{s^2} \mathsf{L}[Ru_2(x,t) + H_2(u)] \right],$$

and so on.

(d) Homotopy Analysis Method (HAM)

Nonlinear equations are difficult to solve, especially ordinary differential equations, partial differential equations, differential–integral equations, differential–difference equations, and coupled differential equations. Unlike perturbation methods, the HAM is independent of small/large physical parameters and provides us a simple way to ensure the convergence of solution series. The technique was first devised by Shijun Liao in 1992; then, more and more researchers have successfully applied this method to various nonlinear problems in science and engineering.

1.6.1   Basic Idea of Homotopy Analysis Method (HAM)

In this section, the basic ideas of the HAM are presented. Here a description of the method is given to handle the general nonlinear problem.

$$N[u(t)] = 0, \quad t > 0, \tag{1.14}$$

where N is a nonlinear operator and $u_0(t)$ is an unknown function of the independent variable t.

1.6.2   Zero-Order Deformation Equation

Let $u_0(t)$ denote the initial guess of the exact solution of Eq. (1.14), $h \neq \mathbf{0}$ is an auxiliary parameter, $H(t) \neq \mathbf{0}$ is an auxiliary function, and L is an auxiliary linear operator with the property

$$L(f(t)) = \mathbf{0}, \; f(t) = \mathbf{0}. \tag{1.15}$$

The auxiliary parameter h, the auxiliary function (t), and the auxiliary linear operator L play an important role within the HAM to adjust and control the convergence region of solution series. Liao constructs, using $q \in [\mathbf{0}, \mathbf{1}]$ as an embedding parameter, the so-called zero-order deformation equation.

$$(\mathbf{1} - q)L[(\emptyset(t; q) - u_0(t)] = qhH(t)N[(\emptyset(t; q)], \tag{1.16}$$

where $\emptyset(t; q)$ is the solution which depends on h, $H(t)$, L, $u_0(t)$, and q. When $q = 0$, the zero-order deformation Eq. (1.14) becomes

$$\emptyset(t; \mathbf{0}) = u_0(t), \tag{1.17}$$

and when $q = 1$, since $h \neq \mathbf{0}$ and $H(t) \neq \mathbf{0}$, the zero-order deformation Eq. (1.14) reduces to

$$N[\emptyset(t; \mathbf{1})] = \mathbf{0}. \tag{1.18}$$

So, $\emptyset(t; \mathbf{1})$ is exactly the solution of the nonlinear equation. Define the so-called mth-order deformation derivatives.

$$u_m(t) = \frac{1}{m!} \frac{\partial^m \emptyset(t; q)}{\partial q^m}. \tag{1.19}$$

If the power series Eq. (1.14) of $\emptyset(t; q)$ converges at $q = 1$; then, we get the following series solution:

$$u(t) = u_0(t) + \sum_{m=1}^{\infty} u_m(t). \tag{1.20}$$

where the terms $u_m(t)$ can be determined by the so-called high-order deformation described below.

1.6.3   Higher-Order Deformation Equation

Define the vector,

$$\overrightarrow{u_n} = \{u_0(t), u_1(t), u_2(t), \ldots, u_n(t)\}. \tag{1.21}$$

Differentiating Eq. (1.16) m-times for embedding parameter q, then setting $q = 0$, and dividing them by factorial of m, we have the so-called mth-order deformation equation.

$$L[u_m(t) - \chi_m u_{m-1}(t)] = hH(t)R_m\left(\overrightarrow{u_m}, t\right), \tag{1.22}$$

where

$$\chi_m = \begin{cases} 0, & m \leq 1 \\ 1, & \text{otherwise} \end{cases} \tag{1.23}$$

and

$$R_m\left(\overrightarrow{u_m}, t\right) = \frac{1}{(m-1)!} \frac{\partial^{m-1} N[\emptyset(t; q)]}{\partial q^{m-1}}. \tag{1.24}$$

For any given nonlinear operator N, the term $R_m\left(\overrightarrow{u_m}, t\right)$ can be easily expressed by Eq. (1.24). Thus, we can gain $u_1(t), u_2(t) \ldots$ using solving the linear high-order deformation with one after the other order in order. The mth-order approximation of $u(t)$ is given by

$$u(t) = \sum_{k=0}^{m} u_k(t). \tag{1.25}$$

ADM, VIM, and HPM are special cases of HAM. When we set $h = -1$ and $H(r, t) = 1$, we will get the same solutions for all the problems by above methods when we set $h = -1$ and $H(r, t) = 1$. When the base functions are introduced, the $H(r, t) = 1$ is properly chosen using the rule of solution expression, a rule of the coefficient of ergodicity, and rule of solution existence.

(e) Differential Transform Method (DTM)

A variety of numerical and analytical methods have been developed to obtain accurate approximate and analytic solutions for the problems in the literature. The classical Taylor series method is one of the earliest analytic techniques to many problems, but it requires a lot of symbolic calculation for the derivatives of functions and the higher-order derivatives. The updated version of the Taylor series is called differential transform method. The DTM is very efficient and robust solver for solving various kinds of differential equations. The main advantage of this technique is that it can be applied directly to linear and nonlinear differential equations without requiring linearization, discretization, or perturbation.

1.6.4   A Few Numerical Examples (Chebyshev Wavelet Method for Solving Reaction–Diffusion Equations (RDEs))

(i) Consider the steady-state RDE

$$\frac{\partial^2 u}{\partial x^2} - \frac{ku}{1 + \alpha u + \beta u^2} = 0 \tag{1.26}$$

with the initial condition
$u(1) = 1,\ u'(0) = 0.$

Solving the Eq. (1.26) using the algorithm for the case corresponds to $M = 2$, $k = 0$ to obtain an approximate solution of $u(x)$. First, if we make use of the two operational matrices, D and D^2 are given, respectively, by

$$D = \begin{pmatrix} 0 & 0 & 0 \\ 4 & 0 & 0 \\ 0 & 8 & 0 \end{pmatrix} \quad D^2 = \begin{pmatrix} 0 & 0 & 0 \\ 0 & 0 & 0 \\ 32 & 0 & 0 \end{pmatrix}$$

$$\psi(x) = \sqrt{\frac{2}{\pi}} \begin{pmatrix} 2 \\ 8x - 4 \\ 32x^2 - 32x + 6 \end{pmatrix}$$

$$C = \begin{pmatrix} c_0 & c_1 & c_2 \end{pmatrix}^{\mathrm{T}} = \sqrt{\frac{\pi}{2}} \begin{pmatrix} c_0 & c_1 & c_2 \end{pmatrix}^{\mathrm{T}}$$

Consider the Eq. (1.26) with $k = 0.1$, $\alpha = 0$ and $\beta = 0$.

$$u''(x) - (0.1)u(x) = 0 \tag{1.27}$$

$$C^T D^2 \psi(x) - (0.1)C^T \psi(x) = 0$$

which is equivalent to

$$64c_2 - (0.1)(2c_0 - 2\sqrt{2}c_1 + 2c_2) = 0$$

$$64c_2 - 0.2c_0 + 0.283c_1 - 0.2c_2 = 0$$

Using the initial condition

$$2c_0 + 4c_1 + 6c_2 = 1$$

$$c_1 = 4c_2$$

Solving the equations, we get

$$c_0 = 0.489, \ c_1 = 0.004, \ c_2 = 0.001$$

Consequently,

$$u(x) = \begin{pmatrix} 0.489 & 0.004 & 0.001 \end{pmatrix} \begin{pmatrix} 2 \\ 8x - 4 \\ 32x^2 - 32x + 6 \end{pmatrix}$$

$$= 0.032x^2 + 0.968$$

(ii) Consider the Eq. (1.26) with $k = 1$, $\alpha = 0$ and $\beta = 0$

$$u''(x) - u(x) = 0$$

The above equation can be written as

$$C^T D^2 \psi(x) - C^T \psi(x) = 0$$

$$64c_2 - (2c_0 - 2\sqrt{2}c_1 + 2c_2) = 0$$

$$62c_2 - 2c_0 + 2\sqrt{2}c_1 = 0$$

Using the initial condition

$$2c_0 + 4c_1 + 6c_2 = 1$$

$$c_1 = 4c_2$$

Solving the equations, we get

$$c_0 = 0.291, c_1 = 0.076 \, c_2 = 0.019$$

Consequently,

$$u(x) = \begin{pmatrix} 0.291 & 0.076 & 0.019 \end{pmatrix} \begin{pmatrix} 2 \\ 8x - 4 \\ 32x^2 - 32x + 6 \end{pmatrix}.$$

$$= 0.608x^2 + 0.392$$

Figure 1.1 shows the comparison between the CWM and VIM for various parameter values.

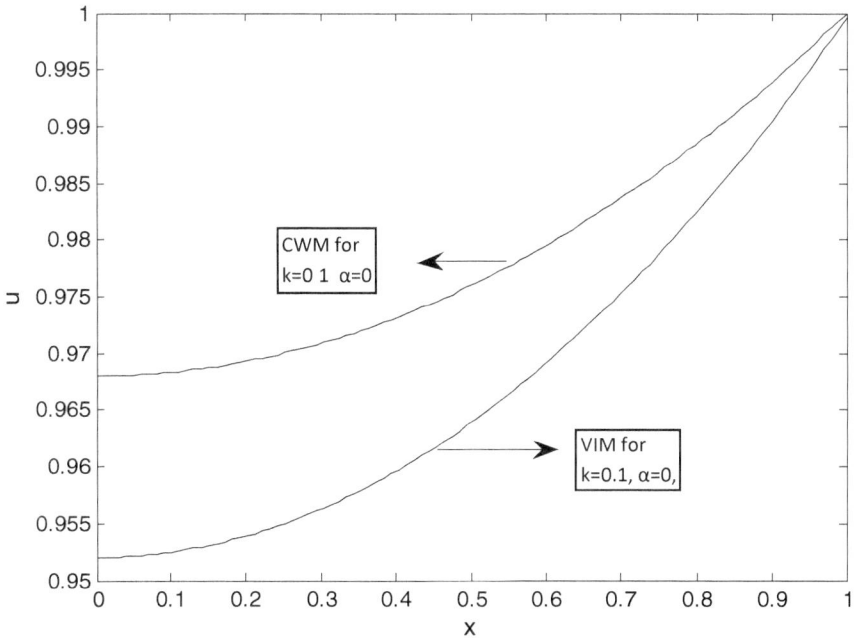

Fig. 1.1 Comparison between CWM and VIM for $k = 0.1$, $\alpha = 0$, and $\beta = 0$

References

1. S. Chandrasekhar, *An Introduction to the Study of Stellar Structure* (Dover Publications, Clayton, 1967)
2. J.H. Lane, On the theoretical temperature of the sun under the hypothesis of a gaseous mass maintaining its internal heat and depending on the laws of gases known to terrestrial experiment, Am. J. Sci. Arts. **2**, 57–74 (1870)
3. H.T. Davis, *Introduction to Nonlinear Differential and Integral Equations* (Dover, New York, 1962)
4. E.H. Doha, W.M. Abd- Elhameed, Y.H. Youssri, Second kind Chebyshev operational matrix algorithm for solving differential equations of Lane-Emden type. New Astron. **23–24**, 113–117 (2013)
5. G. Hariharan, An efficient wavelet based approximation method to water quality assessment model in a uniform channel. Ains Shams Eng. J. (2013 in press)
6. G. Hariharan, K. Kannan, K.R. Sharma, Haar wavelet in estimating the depth profile of soil temperature. Appl. Math. Comput. **210**, 119–225 (2009)
7. G. Hariharan, K. Kannan, Haar wavelet method for solving Fisher's equation. Appl. Math. Comput. **211**, 284–292 (2009)
8. G. Hariharan, K. Kannan, A comparative study of a Haar wavelet method and a restrictive Taylor's series method for solving convection-diffusion equations. Int. J. Comput. Methods Eng. Sci. Mech. **11**(4), 173–184 (2010)
9. G. Hariharan, K. Kannan, Review of wavelet methods for the solution of reaction–diffusion problems in science and engineering. Appl. Math. Model. **38**(1)7, 99–813 (2014)
10. D. Sathiyaseelan, M.B. Gumpu, N. Nesakumar, J.B.B. Rayappan, G. Hariharan, Wavelet based spectral approach for solving surface coverage model in an electrochemical arsenic sensor—an operational matrix approach. Electrochim. Acta **266**, 27–33 (2018)
11. M. Mahalakshmi, G. Hariharan, A new spectral approach on steady-state concentration of species in porous catalysts using wavelets. J. Membr. Biol. **250**, 163–169 (2017)
12. G. Hariharan, D. Sathyaseelan, Efficient spectral methods for a class of unsteady-state free-surface ship models using wavelets. Z. Angew. Math. Phys. **68**, 31 (2017)
13. M.H. Heydari, M.R. Hooshmandas, C. Cattani, G. Hariharan, An optimization wavelet method for multi variable-order fractional differential equations. Fundam. Inf. **151**(1–4), 255–273 (2017)
14. M. Salai Mathi Selvi, G. Hariharan, K. Kannan, M.H. Heydari, Two reliable computational methods pertaining to steady state substrate concentration of an immobilized enzyme system. Alexandria Eng. J. (2017 in press)
15. M. Salai Mathi Selvi, G. Hariharan, K. Kannan, A reliable spectral method to reaction-diffusion equations in entrapped-cell photobioreactor packed with gel granules using Chebyshev wavelets. J. Membr. Biol. **250**(6, 1), 663–670 (2017)
16. Vivek Kumar, Mani Mehra, Cubic spline adaptive wavelet scheme to solve singularly perturbed reaction-diffusion problems. Int. J. Wavelets Multiresolut. Inf. Process. **5**(2), 317–331 (2007)
17. U. Lepik, Numerical solution of evolution equations by the Haar wavelet method. Appl. Math. Comput. **185**, 695–704 (2007)
18. U. Lepik, Application of the Haar wavelet transform to solving integral and differential Equations. Proc. Estonian Acad. Sci. Phys. Math. **56**(1), 28–46 (2007)
19. M. Mehra, N.K.-R. Kevlahan, An adaptive wavelet collocation method for the solution of partial differential equations on the sphere. J. Comput. Phys. **227**, 5610–5632 (2008)

Chapter 2
Wavelet Analysis—An Overview

2.1 Wavelet Analysis

Wavelet analysis is a new branch of mathematics widely applied in signal analysis, image processing, numerical analysis, etc. The name wavelets means small waves (the sinusoids used in Fourier analysis are 'big waves'), and in short, a wavelet is an oscillation that decays quickly.

Theory of wavelets has been employed over the past decades. Grossmann and Morlet, a physicist and an engineer, introduced the word wavelet for the first time. Wavelet means small wave. So wavelet analysis is about analyzing a mathematical function (in engineering it is a signal) with short duration finite energy functions. Then the wavelet transforms decompose a signal into a set of frequency referred to as scales by projecting the signal onto an element of a set of basis functions. These basis functions are called wavelets, varying by dilation and translation. In the last decade, many mathematicians developed their ideas in mathematical thinking and applied these ideas to applied mathematics, such as finding the numerical solutions to several kinds of differential equations, using them in signal processing, image analysis, and many other fields.

Wavelets are used to transform the given mathematical function (signal) under investigation into another representation which is in more useful form to analyze. This transformation of the mathematical function (signal) is called wavelet transform.

In this book, we will show how wavelets can be used to solve reaction–diffusion equations, which are characterized by widely varying length scales and which are therefore hardly accessible by other numerical methods. In most papers, the Daubechies wavelets are applied. These wavelets are orthogonal and sufficiently smooth and have a compact support. Their shortcoming is that an explicit expression is lacking. This obstacle makes the differentiation and integration of these wavelets very complicated. Among the wavelet families, which are defined by an analytical expression, special attention deserves the Haar wavelets. In 1910,

© Springer Nature Singapore Pte Ltd. 2019
G. Hariharan, *Wavelet Solutions for Reaction–Diffusion Problems in Science and Engineering*, Forum for Interdisciplinary Mathematics,
https://doi.org/10.1007/978-981-32-9960-3_2

Alfred Haar introduced the notion of wavelets. His initial theory has been expanded recently into a wide variety of applications, but primarily it allows for the representation of various functions by a combination of step functions and wavelets over specified interval widths.

The oldest and probably best-known method for this is the Fourier transform developed in 1807 by Joseph Fourier. An alternative method with some attractive properties is the wavelet transform, first mentioned by Alfred Haar in 1909. Since then a lot of researches into wavelets and the wavelet transform are performed. However, the trigonometric analysis functions are replaced by a wavelet function. A wavelet is a short oscillating function, which contains both the analysis function and the window. Time information is obtained by shifting the wavelet over the signal.

The wavelet transform has been perhaps the most exciting development in the decade to bring together researchers in several different field such as signal processing, quantum mechanics, image processing, communications, computer science, and mathematics—to name a few. Today, wavelet is not only the workspace in computer imaging and animation; they are also used by the FBI to encode its database of million fingerprints. In future, scientist may put wavelet analysis for diagnosing breast cancer, looking for heart abnormalities, predicting the weather, signal processing, data compression, smoothing and image compression, fingerprints verification, DNA analysis, protein analysis, blood pressure, heart rate and ECG analysis, finance, Internet traffic description, speech recognition, and computer graphics.

2.2 Comparison Between Fourier Transform (FT) and Wavelet Transform (WT)

In Fourier-based methods, since the products of the basis elements are also basis elements, the procedure does not face any difficulty. The interaction between wavelets and the heat equation has been made possible by first applying wavelet decomposition to the initial state and then investigating the evolution of each wavelet component. Another channel that has connected the heat equation and wavelets is their common probabilistic background. The phenomenon of heat diffusion is closely related to Gaussian random process (through the kernel function). Similarly, the fundamental equation in wavelet theory, the refinement equation, also allows probabilistic interpretations (Derfel and Shen). As a result, we are able to view heatlets or related functions in the framework of probability theory. This idea points to a computational algorithm for heatlets.

This sparseness, in turn, results in a number of useful applications such as data compression, detecting features in images, and removing noise from time series. One way to see the time-frequency resolution differences between the Fourier transform, and the wavelet transform is to look at the basis function coverage of the time-frequency plane. An advantage of wavelet transforms is that the windows vary. In order to isolate signal discontinuities, one would like to have some very short basic functions. At the same time, in order to obtain detailed frequency analysis, one would like to have some very long basis functions. A way to achieve this is to have short high-frequency basis functions and long low-frequency ones. This happy medium is exactly what you get with wavelet transforms. One thing to remember is that wavelet transforms do not have a single set of basis functions like the Fourier transform, which utilizes just the sine and cosine functions. Instead, wavelet transforms have an infinite set of possible basis functions. Thus, wavelet analysis provides immediate access to information that can be obscured by other time-frequency methods such as Fourier analysis.

2.3   Wavelets and Multi-resolution Analysis (MRA)

This chapter provides a brief introduction to wavelets and multi-resolution analysis (MRA). Also it describes applications of wavelet methods for solving differential equations.

Morlet and Grossmann introduced the concept of wavelets in early 1980s. Since then, a lot of researchers were involved in the development of wavelets theory. A few notable contributors include Morlet and Grossmann for mathematical formulation of continuous wavelet transform (CWT). Mayer and Mallat used MRA using wavelet transforms, and Daubechies proposed the orthogonal compactly supported wavelets. More recently, wavelet-based approximation methods have been successfully utilized for the solutions of ordinary differential, integral, integro-differential, and partial differential equations. Several methods such as wavelet-Galerkin, wavelet-collocation, Taylor-wavelet-Galerkin have been studied both from the theoretical and the computational points of view. Multi-resolution analysis (MRA) gives a very good understanding of the structure of wavelets. Wavelet analysis is also more flexible in the sense that one can choose a specific wavelet to match the type of function being analyzed. There are many kinds of wavelets with different properties available in literature. Some examples are the Haar wavelets [1–3]. Haar wavelet is the simplest orthonormal compact support wavelet. It is also called mother wavelet. Some nonlinear parabolic differential equations are solved by wavelet method.

2.4   Evolution of Wavelets

In Fourier transform (FT), the function $f(t)$ is a periodic function which can be expressed as an infinite sum of complex exponentials. Infinite FT pair is defined as

$$F(f(t)) = F(s) = \int_{-\infty}^{\infty} f(t)e^{ist}dt, \qquad (2.1)$$

$$f(t) = F^{-1}(F(s)) = \int_{-\infty}^{\infty} F(s)e^{-ist}ds. \qquad (2.2)$$

Using above equations, a time domain signal $f(t)$ can be transformed into the frequency domain and back again.

Discrete Fourier transform (DFT) pair is defined as

$$F_n = F_{\text{DFT}}(f) = \frac{1}{N}\sum_{k=0}^{N-1} f_k e^{-is\Delta t}. \qquad (2.3)$$

The inverse discrete Fourier transform is

$$f_k = \frac{1}{N}\sum_{n=}^{N-1} F_n e^{-is\Delta t}. \qquad (2.4)$$

The computation of DFT is very time-consuming for large signals (large N). The fast Fourier transform (FFT) does not take an arbitrary number of intervals N but only the ranges $N = 2^m$, $m \in N$. By reducing the number of intervals, the FFT becomes very fast. A significant drawback is that the signal should have 2^m samples.

The shift from the Fourier transform to the wavelet transform is best explained through short-time Fourier transform (STFT) or windowed Fourier transform (WFT). The STFT calculates the Fourier transform of windowed part of the signal where the window shifts along the time axis. The limitation of the Fourier transforms which gives only frequency content of the signal is overcome by STFT. STFT can retrieve both frequency and time information from the signal

$$\text{STFT}_f(s, \tau) = \int_{-\infty}^{\infty} f(t)g^*(t - \tau)e^{-ist}dt. \qquad (2.5)$$

The performance of STFT depends on the chosen window $g(t)$. A short window gives a good time resolution, but different frequencies are not identified very well. A long window gives an inferior time resolution but a better frequency resolution. But finding the proper window to get both time and frequency resolutions is critical for the quality of STFT.

2.5   Genesis of Wavelets

- Joseph Fourier (1807) designed a method for representing a signal with a series of coefficients based on trigonometric functions.
- Alfred Haar (1909) established the Haar basis function with compact support (vanish outside finite interval).
- In 1980, Alex Grossmann (physicist) and Jean Morlet (Engineer) derived a decomposing signal method into wavelet coefficients.
- Stephane Mallat (1985) developed a multi-resolution analysis (MRA) using wavelets.
- Ingrid Daubechies (1988) used Mallat work to construct her own family of wavelets.

2.6   Continuous-Time Wavelets

Here we consider a real- or complex-valued continuous-time function $\psi(t)$ with the following properties.

1. The function integrates to zero.

$$\int_{-\infty}^{\infty} \psi(t)dt = 0. \tag{2.6}$$

2. It is square integrable or equivalently has finite energy.

$$\int_{-\infty}^{\infty} |\psi(t)|^2 dt < \infty. \tag{2.7}$$

3. Wavelet admissibility condition

$$c_\psi = \int_0^{\infty} \frac{|\hat{\psi}(\omega)|}{\omega} d\omega < \infty, \tag{2.8}$$

where $\hat{\psi}(\omega)$ is the Fourier transform of $\psi(t)$.

It implies that the wavelet should have no zero-frequency components. Then the function $\psi(t)$ is called a mother wavelet. Entire wavelet has finite duration; then it is supported compactly

$$0 \leq t \leq c.$$

The mother wavelet is defined by

$$\psi_{a,b}(t) = |a|^{-\frac{1}{2}}\psi\left(\frac{t-b}{a}\right), \quad a, b \in R, \quad a \neq 0. \tag{2.9}$$

The mother wavelet is dilated and translated to form a basis for Hilbert space $L^2(R)$. Here b is a position (translation) parameter, and a is a scaling (dilation) parameter. $|a|^{-\frac{1}{2}}$ is a normalizing constant.

The continuous wavelet transform is given by

$$w(a,b) = |a|^{-\frac{1}{2}} \int\limits_{-\infty}^{\infty} f(t)\psi\left(\frac{t-b}{a}\right)dt, \quad a, b \in R, \quad a \neq 0. \tag{2.10}$$

The elements in $w(a, b)$ are called wavelet coefficients.

2.7 Discrete Wavelet Transform (DWT)

Due to the redundancy of continuous wavelet transform, it should be possible to discretize the dilation and translation variables and still obtain a stable reconstruction. Furthermore, it is possible to get a real orthonormal basis. A logarithmic discretization is chosen for scaling parameter 'a' i.e., $a = a_0^m$ with m being an integer and a_0 is not equal to 1. For the translation parameter 'b', it is chosen $b = nb_0a_0^m$ with m and n are integers. Then $\psi(a_0^m t - nb_0)$ covers the whole time axis for any given scale a_0^m.

Now the family of discrete wavelet functions is given by

$$\psi_{m,n} = |a_0|^{\frac{m}{2}}\psi(a_0^m t - nb_0), \quad m, n \in Z. \tag{2.11}$$

2.8 Desirable Properties of Wavelets

In this section, some general features of wavelets are discussed.

1. **Compact support**

The wavelet function $\psi_{m,n}(t)$ has a finite compact support when it vanishes after a finite interval, i.e., $0 < t < k$ (k is a constant). Then the coefficients in the dilation equation are finite. This property is useful in solving differential equations.

2. **Orthogonality and Orthonormality**

Two real functions $f_1(t)$ and $f_2(t)$ are said to be orthonormal if and only if $\int_{-\infty}^{\infty} f_1(t) f_2(t) dt = 0$ (Orthogonality), and

$$\int_{-\infty}^{\infty} f_i(t) f_i(t) = 1 \text{ for } i = 1, 2, \ldots \tag{2.12}$$

3. **Dyadic**

Generally, for the discrete wavelet function ψ_{nm} is $|a_0|^{\frac{m}{2}} \psi(a_0^m t - nb_0)$, m, $n \in Z$.

It is fine to choose translation parameter a_0 sufficiently close to 1, and dilation parameter b_0 is close to 0. However, for $a_0 = 2$, $b_0 = 1$; the dyadic case is obtained for which it is known that orthonormal basis exists, and reconstruction from the transformed wavelet coefficients is possible.

2.9 Multi-resolution Analysis (MRA)

An orthogonal multi-resolution analysis is a decomposition of Hilbert space $L^2(R)$ into a chain of nested closed subspaces such that

1. $\ldots \subset V_{N-2} \subset V_{N-1} \subset V_0 \subset V_1 \ldots \subset L^2(R)$ (nestedness)
2. $\cap_{j \in z} V_j = \{0\}$, and $\cup_{j \in z} V_j$ is dense in $L^2(R)$ (completeness)
3. $f(t) \in V_j$ iff $f(2t) \in V_{j+1}$ (scaling property)
4. $f(t) \in V_0$ iff $f(t - k) \in V_0$ (translation invariance)
5. There exists a scaling function $\phi \in V_0$ such that $\{\phi(t - k)\}$ is an orthonormal basis of V_0.

The chain of subspaces in an MRA is completely defined by its scaling function φ. The initial subspace V_0 is the closure of the span of the integer translates. To create the other subspaces, we use scaling property that allows moving upwards and downwards on the scale of subspaces.

We can define the scaling function as

$$\varphi_{j,k}(t) = 2^{\frac{j}{2}} \varphi(2^j t - k), \tag{2.13}$$

Here $\varphi_{j,k} \in V_j$, $\varphi_{j,k}$ is the orthonormal basis of V_j.

2.10   Discrete Wavelet Transforms Methods

2.10.1   Haar Wavelets

Haar wavelet is a system of square wave; the first curve was marked up as $h_0(t)$; the second curve marked up as $h_1(t)$ that is

$$h_0(t) = \begin{cases} 1, & 0 \leq x < 1 \\ 0, & \text{otherwise} \end{cases} \tag{2.14}$$

$$h_1(t) = \begin{cases} 1, & 0 \leq x < 1/2, \\ -1, & 1/2 \leq x < 1, \\ 0, & \text{otherwise,} \end{cases} \tag{2.15}$$

where $h_0(t)$ is scaling function and $h_1(t)$ is mother wavelet. To perform wavelet transform, Haar wavelet uses dilations and translations of a function, i.e., the transform makes the following function.

$$h_n(t) = h_1\left(2^j t - k\right), \quad n = 2^j + k, \ j \geq 0, \ 0 \leq k < 2^j. \tag{2.16}$$

Chen and Hsiao [4] described the ideology of operational matrix in 1992 and investigated the generalized integral operational matrix, that is, the integral of a matrix $\phi(t)$ can be approximated as follows:

$$\int_0^t \phi(t)\mathrm{d}t \cong Q_\phi \phi(t), \tag{2.17}$$

where Q_ϕ is an operational matrix of a one-time integral matrix $\phi(t)$; similarly, we can get an operational matrix Q_ϕ^n of the n-time integral of $\phi(t)$. For example, the operational matrix of $\Phi(t)$ can be expressed by following:

$$Q_\Phi = \Phi Q_B \Phi^{-1}. \tag{2.18}$$

Here Q_B is the operational matrix of the block pulse function:

$$Q_{B_m} = \frac{1}{2m} \begin{bmatrix} 1 & 2 & 2 & \cdots & 2 \\ 0 & 1 & 2 & \cdots & 2 \\ 0 & 0 & 1 & \cdots & 2 \\ \vdots & \vdots & \vdots & \vdots & \vdots \\ 0 & 0 & 0 & 0 & 1 \end{bmatrix},$$

where m is the dimension of the matrix $\Phi(t)$, and usually, $m = 2^\alpha$, α is a positive integer.

If $\Phi(t)$ is a unitary matrix, then $Q_\Phi = \Phi Q_B \Phi^{\mathrm{T}}$, Q_Φ is a matrix with the characteristic of briefness and profound utility.

For $x \in [0, 1]$, Haar wavelet function is defined as

$$h_0(x) = \frac{1}{\sqrt{m}}. \tag{2.19}$$

$$h_i(x) = \frac{1}{\sqrt{m}} \begin{cases} 2^{\frac{j}{2}}, & \frac{k-1}{2^j} \le x < \frac{k-(1/2)}{2^j} \\ -2^{\frac{j}{2}}, & \frac{k-(1/2)}{2^j} \le x < \frac{k}{2^j} \\ 0, & \text{otherwise} \end{cases} \tag{2.20}$$

Integer $m = 2^j$ $(j = 0, 1, 2, \ldots, J)$ indicates the level of the wavelet; $i = 0, 1, 2, \ldots, m - 1$ is the translation parameter. The maximal level of resolution is J. The index i is calculated according to the formula $i = m + k - 1$; in the case of minimal values $m = 1$, $k = 0$, we have $i = 2$, the maximal value of i is $i = 2M = 2^{J+1}$. It is assumed that the value $i = 1$ corresponds to the scaling function for which $h_1 \equiv 1$ in $[0, 1]$. Let us define the collocation points $t_l = (l - 0.5)/2M$, $(l = 1, 2, \ldots, 2M)$ and discretize the Haar function $h_i(x)$; in this way, we get the coefficient matrix $H(i, l) = (h_i(x_l))$, which has the dimension $2M \times 2M$.

2.10.2   Function Approximation

Any square-integrable function $y(x, t) \in L^2[0, 1)$ defined in the interval $0 \le x \le 1$ and $0 \le t \le 1$ can be expanded by a Haar series of infinite terms

$$y(x, t) \approx \sum_{i=0}^{m-1} \sum_{j=0}^{m-1} c_{i,j} h_i(x) h_j(t), \tag{2.21}$$

where the Haar coefficients c_{ij} are determined as

$$c_{i,j} = \int_0^1 h_j(x)\mathrm{d}x. \int_0^1 y(x, t)h_j(t)\mathrm{d}t, \quad (i, j = 0, 1, 2, \ldots, m - 1). \tag{2.22}$$

Equation (2.21) can be written into the discrete form by

$$Y(x, t) = H^{\mathrm{T}}(x)CH(t). \tag{2.23}$$

Now $\overrightarrow{h_0^{\mathrm{T}}} = [h_{0,0} \quad h_{0,1} \quad \cdots \quad h_{0,m-1}]$, $\overrightarrow{h_1^{\mathrm{T}}} = [h_{1,0} \quad h_{1,1} \quad \cdots \quad h_{1,m-1}], \ldots, = [h_{m-1,0} \quad h_{m-1,1} \quad \cdots \quad h_{m-1,m-1}]$ are the discrete form of the Haar wavelet bases;

the discrete values are taken from the continuous curves $h_0(t), h_1(t), \ldots, h_{m-1}(t)$, respectively. The Haar wavelet matrix H of dimension m is defined by

$$H = \begin{bmatrix} \overrightarrow{h_0^T} \\ \overrightarrow{h_1^T} \\ \vdots \\ \overrightarrow{h_{m-1}^T} \end{bmatrix} = \begin{bmatrix} h_{0,0} & h_{0,1} & \cdots & h_{0,m-1} \\ h_{1,0} & h_{1,1} & \cdots & h_{1,m-1} \\ \vdots & \vdots & \vdots & \vdots \\ h_{m-1,0} & h_{m-1,1} & \cdots & h_{m-1,m-1} \end{bmatrix}. \tag{2.24}$$

$$C = \begin{bmatrix} c_{0,0} & c_{0,1} & \cdots & c_{0,m-1} \\ c_{1,0} & c_{1,1} & \cdots & c_{1,m-1} \\ \vdots & \vdots & \vdots & \vdots \\ c_{m-1,0} & c_{m-1,1} & \cdots & c_{m-1,m-1} \end{bmatrix}, \tag{2.25}$$

C is the coefficient matrix of Y, and it can be obtained by the formula:

$$C = H \cdot Y \cdot H^T. \tag{2.26}$$

For deriving the operational matrix of Haar wavelets,

$$QH = H \cdot Q_B \cdot H^T, \tag{2.27}$$

where $Q_B = \frac{1}{m} \begin{bmatrix} \frac{1}{2} & 1 & \cdots & 1 \\ 0 & \cdots & \cdots & \vdots \\ \vdots & 0 & \frac{1}{2} & 1 \\ 0 & \cdots & 0 & \frac{1}{2} \end{bmatrix}$ is the operational matrix of the block pulse functions.

Chen and Hsiao [4] showed that the following matrix equation for calculating the matrix P of order m holds

$$P_{(m)} = \frac{1}{2m} \begin{pmatrix} 2mP_{(m/2)} & -H_{(m/2)} \\ H_{(m/2)}^{-1} & O \end{pmatrix},$$

where O is a null matrix of order $\frac{m}{2} \times \frac{m}{2}$.

The operational matrix of integration P, which is a $2M$ square matrix, is defined by the equation

$$PH(t) \cong \int_0^t H(t)\mathrm{d}t. \tag{2.28}$$

$$QH(t) \cong \int_0^t PH(t)\mathrm{d}t. \tag{2.29}$$

$$P_1 = \begin{bmatrix} \dfrac{1}{2} \end{bmatrix},$$

$$H_2 = \begin{pmatrix} 1 & 1 \\ 1 & -1 \end{pmatrix}, \quad P_2 = \frac{1}{4}\begin{pmatrix} 2 & -1 \\ 1 & 0 \end{pmatrix},$$

$$H_4 = \begin{bmatrix} 1 & 1 & 1 & 1 \\ 1 & 1 & -1 & -1 \\ 1 & -1 & 0 & 0 \\ 0 & 0 & 1 & -1 \end{bmatrix}, \quad P_4 = \frac{1}{16}\begin{bmatrix} 8 & -4 & -2 & -2 \\ 4 & 0 & -2 & 2 \\ 1 & 1 & 0 & 0 \\ 1 & -1 & 0 & 0 \end{bmatrix},$$

$$P_8 = \frac{1}{64}\begin{bmatrix} 32 & -16 & -8 & -8 & -4 & -4 & -4 & -4 \\ 16 & 0 & -8 & 8 & -4 & -4 & 4 & 4 \\ 4 & 4 & 0 & 0 & -4 & 4 & 0 & 0 \\ 4 & 4 & 0 & 0 & -4 & 4 & 0 & 0 \\ 1 & 1 & 2 & 0 & 0 & 0 & 0 & 0 \\ 1 & 1 & -2 & 0 & 0 & 0 & 0 & 0 \\ 1 & -1 & 0 & 2 & 0 & 0 & 0 & 0 \\ 1 & -1 & 0 & -2 & 0 & 0 & 0 & 0 \end{bmatrix}$$

$$H_{m \times m} \triangleq [h_m(t_0) \quad h_m(t_1) \ldots h_m(t_{m-1})], \tag{2.30}$$

here $\frac{i}{m} \le t < \frac{i+1}{m}$.
and

$$H_{m \times m}^{-1} = \frac{1}{m} H_{m \times m}^{\mathrm{T}} \, rIm,$$

$$r = \begin{bmatrix} 1, \ 1, \ 2, \ 2, \ 4, \ 4, \ 4, \ 4, \ \ldots, \ \underbrace{\frac{m}{2}, \frac{m}{2}, \ldots, \frac{m}{2}}_{\frac{m}{2}\text{elements}} \end{bmatrix}^{\mathrm{T}}, \quad m > 2 \tag{2.31}$$

It should be noted that calculations for $P_{(m)}$ and $H_{(m)}$ must be carried out only once; after that, they will be applicable for solving whatever differential equations.

2.11 Wavelet Method for Solving a Few Reaction–Diffusion Problems—Status and Achievements

Wavelets have been applied recently to obtain representations of integral and differential operators in many physical problems. The first applications of wavelets to the solution of PDEs seem to have consisted of Galerkin methods on problems with periodic boundary conditions as shown in [5]. In 1992, Chen and Hsiao [4] introduced the Haar wavelet method for solving one-dimensional diffusion equation. Cattani [6, 7] started to publish wavelet solution of diffusion–hyperbolic equations since 2002, by using harmonic wavelets, Shannon wavelets, and connection coefficients, but also by using some regularized connection coefficients for the Haar wavelets. Celik [8] used the Haar wavelet method for solving the Fisher's equation, and he showed that the Haar wavelet method is convenient tool for nonlinear partial differential equations.

In the area differential operators, most of the attempts of using wavelets is to generate an adaptive meshing structure (select a partial set of full wavelet basis functions) upon which differential operations can be carried out, thus reducing the amount of mesh points needed to resolve detailed structures in the solutions of PDEs [5–17]. The application of wavelet-based methods to the numerical solution of partial differential equations (PDEs) has recently been studied from both the theoretical and the computational point of view due to its attractive feature: orthogonality, arbitrary regularity, good localization. This algorithm is based on a collocation method for PDEs with a specially designed spline wavelet for the Sobolev space $H^2(I)$ on a closed interval I, and they have addressed that the SW-ADI method is an efficient time-dependent adaptive method with second-order accuracy for solutions with localized phenomena, such as in flame propagations. Issues like new boundary wavelets for more accurate boundary conditions, adaptive strategy for two-dimensional meshes, data structure and storage and implementation details, and numerical results have also been addressed.

Rathish Kumar and Mani Mehra [5, 16, 17] introduced the Taylor-wavelet-Galerkin method for Burger's equation, Korteweg–de Vries equation, and hyperbolic and parabolic problems. Lepik [18] introduced the Haar wavelet method for solving some evolution equations. In this article, he established the Haar wavelet solutions of one-dimensional Burgers' equations and sine-Gordon equations with various boundary and initial conditions. The same author established the segmentation method for solving differential equations and one-dimensional heat equation through Haar wavelet. In his works, the Haar wavelet solutions had compared with other classical solutions. In the recent years, Hariharan and his co-workers [9–11] introduced the wavelet method for solving a few reaction–diffusion problems.

2.11.1 Importance of Operational Matrix Methods for Solving Reaction–Diffusion Equations

The main characteristic of these approaches is that in the fundamental matrix approaches there is not any approximating symbol. An error estimation of the collocation scheme for solving the generalized nonlinear differential equation such as reaction diffusion equation can be noted. The rate of convergence of numerical methods due to operational matrices is exponential in simple geometry. Operational matrices of differentiations have several specific properties, fundamental relations which are based on differentiation matrices. By means of the matrix relations between the respective polynomials and their derivatives, an operational matrix derivative is developed together with a set of collocation nodes. The approximation of the solution together with these collocation nodes is utilized to reduce the computation of the problem to some algebraic equations. Also the method is computationally attractive.

Reaction–diffusion equations in many cases can develop very steep solution gradients, even exhibit finite time singularity development or 'blow up.' Wavelet methods would be very suitable for such problems in the time-dependent case, in a scheme in which one adaptively tunes the local wavelet resolution based on the current solution properties.

A Few Examples (Haar Wavelet Method (HWM) for Solving RDEs)
Consider the Fisher's equation

$$\frac{\partial u}{\partial t} = \frac{\partial^2 u}{\partial x^2} + u(1 - u) \tag{2.32}$$

with the initial condition $u(x, 0) = f(x)$, $0 \leq x \leq 1$ and the boundary conditions $u(0, t) = g_0(t)$, $u(1, t) = g_1(t)$, $0 < t \leq T$.
Let us divide the interval $(0, 1]$ into N equal parts of length $\Delta t = (0, 1]/N$ and denote $t_s = (s - 1)\Delta t$, $s = 1, 2, \ldots, N$. We assume that $\dot{u}''(x, t)$ can be expanded in terms of Haar wavelets as formula

$$\dot{u}''(x, t) = \sum_{n=0}^{m-1} c_s(n) h_n(x) = c_{(m)}^{\mathrm{T}} h_{(m)}(x) \tag{2.33}$$

where $\cdot$ and $'$ means differentiation with respect to t and x respectively, the row vector $c_{(m)}^{\mathrm{T}}$ is constant in the subinterval $t \in (t_s, t_{s+1}]$.
Integrating Formula (2.33) with respect to t from t_s to t and twice with respect to x from 0 to x, we obtain

$$u''(x,\ t) = (t - t_s)c_{(m)}^{\mathrm{T}}h_{(m)}(x) + u''(x,\ t_s) \tag{2.34}$$

$$u(x,t) = (t - t_s)c_{(m)}^{\mathrm{T}}P_{(m)}^2h_{(m)}(x) + u(x,\ t_s) - u(0,\ t_s) \\ + x[u'(0,\ t) - u'(0,\ t_s)] + u(0,\ t) \tag{2.35}$$

$$\dot{u}(x,t) = c_{(m)}^{\mathrm{T}}P_{(m)}^2h_{(m)}(x) + x\dot{u}'(0,\ t) + \dot{u}(0,\ t) \tag{2.36}$$

By the boundary conditions, we obtain

$$u(0,\ t_s) = g_0(t_s), \quad u(1,\ t_s) = g_1(t_s)$$
$$\dot{u}(0,\ t) = g_0'(t), \quad \dot{u}(1,\ t) = g_1'(t)$$

Putting $x = 1$ in Formulae (2.35) and (2.36), we have

$$u'(0,\ t) - u'(0,\ t_s) = -(t - t_s)c_{(m)}^{\mathrm{T}}P_{(m)}^2h_{(m)}(x) + g_1(t) \\ - g_0(t) - g_1(t_s) + g_0(t_s) \tag{2.37}$$

$$\dot{u}'(0,t) = g_0'(t) - c_{(m)}^{\mathrm{T}}P_{(m)}^2h_{(m)}(x) - g_0'(t) \tag{2.38}$$

Substituting Formulae (2.37) and (2.38) into Formulae (2.34)–(2.36), and discretizing the results by assuming $x \to x_l$, $t \to t_{s+1}$, we obtain

$$u''(x_l, t_{s+1}) = (t_{s+1} - t_s)c_{(m)}^{\mathrm{T}}h_{(m)}(x_l) + u''(x_l, t_s) \tag{2.39}$$

$$u(x_l, t_{s+1}) = (t_{s+1} - t_s)c_{(m)}^{\mathrm{T}}P_{(m)}^2h_{(m)}(x_l) + u(x_l, t_s) - g_0(t_s) + g_0(t_{s+1}) \\ + x_l\left[-(t_{s+1} - t_s)c_{(m)}^{\mathrm{T}}P_{(m)}f + g_1(t_{s+1}) - g_0(t_{s+1}) - g_1(t_s) + g_0(t_s)\right] \tag{2.40}$$

$$\dot{u}(x_l, t_{s+1}) = c_{(m)}^{\mathrm{T}}P_{(m)}^2h_{(m)}(x) + g_0'(t_{s+1}) + x_l\left[-c_{(m)}^{\mathrm{T}}P_{(m)}f + g_1'(t_{s+1}) - g_0'(t_{s+1})\right] \tag{2.41}$$

where the vector f is defined as

$$f = [1, \ \underbrace{0, \ \ldots, \ 0}_{(m-1)\ \text{elements}}]^{\mathrm{T}}$$

In the following, the scheme

$$\dot{u}(x_l, t_{s+1}) = u''(x_l, t_{s+1}) + u(x_l, t_{s+1})(1 - u(x_l, t_{s+1})) \qquad (2.42)$$

which leads us from the time layer t_s to t_{s+1} is used.

Substituting Eqs. (2.39)–(2.41) into the Eq. (2.42), we gain

$$c_{(m)}^T P_{(m)}^2 h_{(m)}(x_l) + x_l \left[-c_{(m)}^T P_{(m)} f + g_1'(t_{s+1}) - g_0'(t_{s+1}) \right] + g_0'(t_{s+1})$$
$$= u''(x_l, t_{s+1}) + u(x_l, t_{s+1})[1 - u(x_l, t_{s+1})] \qquad (2.43)$$

From Formula (2.43), the wavelet coefficients $c_{(m)}^T$ can be successively calculated. Figure 2.1 shows the comparison between the Haar and exaction for $m = 16$.

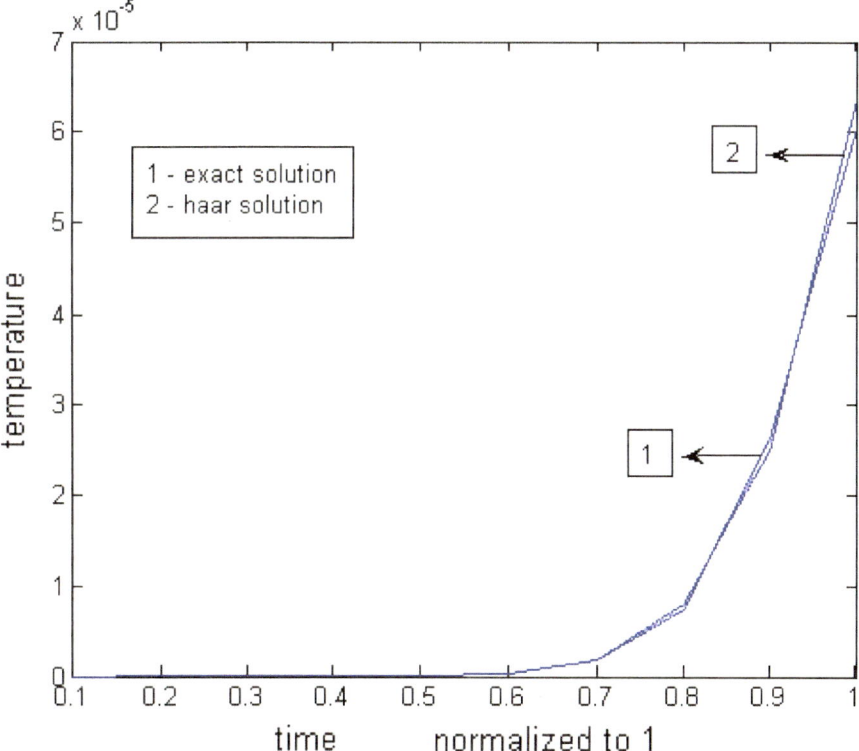

Fig. 2.1 Comparison between exact and Haar solution of Fisher's equation (Problem 2) with $m = 16$

Numerical Experiments
Problem: 1 Consider the following problem

$$\frac{\partial u}{\partial t} = \frac{\partial^2 u}{\partial x^2} + u(1-u), \quad 0<x<1$$

subject to a constant initial condition $u(x,0) = \lambda$.

$$c_{(m)}^T P_{(m)}^2 h_{(m)}(x_l) + x_l\left[-c_{(m)}^T P_{(m)}\lambda + g_1'(t_{s+1}) - g_0'(t_{s+1})\right] + g_0'(t_{s+1})$$
$$= u''(x_l, t_{s+1}) + u(x_l, t_{s+1})[1 - u(x_l, t_{s+1})]$$

(2.44)

From Formula (2.44), the wavelet coefficients $c_{(m)}^T$ can be successively calculated.

Using Adomian decomposition method, the exact solution in a closed form is given by

$u(x,t) = \frac{\lambda e^t}{1-\lambda+\lambda e^t}$ which is in full agreement with the proposed results.

Computer simulation was carried out in the cases $m = 32$ and $m = 64$, the computed results were compared with the exact solution, more accurate results can be obtained by using a larger m.

Problem: 2 In this case, we will examine the Fisher's equation

$$\frac{\partial u}{\partial t} = \frac{\partial^2 u}{\partial x^2} + u(1-u), \quad 0<x<1 \tag{2.45}$$

Subject to the initial condition

$$u(x,0) = \frac{1}{(1+e^x)^2}$$

From Formula (2.43), the wavelet coefficients $c_{(m)}^T$ can be successively calculated.

This process is started with

$$u(x_l, t_s) = \frac{1}{(1+e^x)^2}$$
$$u'(x_l, t_s) = \frac{-2e^x}{(1+e^x)^3}$$
$$u''(x_l, t_s) = -2\left[\frac{e^x - 2e^{2x}}{(1+e^x)^4}\right]$$

References

1. C.K. Chui, *An Introduction to Wavelets* (Academic Press, Boston, 1992)
2. I. Daubechies, *Ten Lectures on Wavelets*, vol. 61 (1992)
3. Y. Meyer (ed.), Wavelets and applications, in *Proceedings of International Conference*. Research Notes in Applied Mathematics, vol. 20, May, 1989 (Springer, Marseille, 1992)
4. C.F. Chen, C.H. Hsiao, Haar wavelet method for solving lumped and distributed-parameter systems. IEEE Proc. Part D **144**(1), 87–94 (1997)
5. B.V. Rathish Kumar, M. Mehra, *Time Accurate Fast Wavelet-Taylor Galerkin Method for Partial Differential Equations*. (Wiley Inter Science, 2009). https://doi.org/10.1002/num. 20092
6. C. Cattani, *Wave Propagation of Shannon Wavelets*. (ICISA, 2006), pp. 7781–7799
7. C. Cattani, A. Kudreyko, *Multiscale Analysis of the Fisher equation, ICCSA 2008, Part I, LNCS 5072* (2008), pp. 1171–1180
8. I. Celik, Haar wavelet method for solving generalized Burgers-Huxley eq. Arab J. Math. Sci **18**, 25–37 (2011)
9. D. Sathiyaseelan, M.B. Gumpu, N. Nesakumar, J.B.B. Rayappan, G. Hariharan, Wavelet based spectral approach for solving surface coverage model in an electrochemical arsenic sensor—an operational matrix approach. Electrochim. Acta **266**, 27–33 (2018)
10. M. Mahalakshmi, G. Hariharan, A new spectral approach on steady-state concentration of species in porous catalysts using wavelets. J. Membr. Biol. **250**, 163 (2017)
11. G. Hariharan, D. Sathyaseelan, Efficient spectral methods for a class of unsteady-state free-surface ship models using wavelets. Z. Angew. Math. Phys. **68**, 31 (2017)
12. S. Padma, G. Hariharan, S. Castellucci, A new spectral method applied to immobilized biocatalyst model arising in biochemical engineering. Contemp. Eng. Sci. **10**(6), 291–304 (2017)
13. C.K. Chui, E. Quak, Wavelets on a bounded interval, in *Numerical Methods of Approximation Theory*, ed. by D. Braess, L.L. Schumaker (Birkhauser, Basel, 1992), pp. 1–24
14. R.R. Coifman, Y. Meyer, M.V. Wickerhauser, Wavelet analysis and signal processing, in *Wavelet Analysis and Signal Processing, in Wavelets and their Applications* (1992), pp. 153–178
15. A. Cohen, Wavelets in numerical analysis, in *The Handbook of Numerical Analysis*, vol. VII, ed. by P.G. Ciarlet, J.L. Lions (Elsevier, Amsterdam, 1999)
16. B.V. Rathish Kumar, M. Mehra, A time-accurate pseudo-wavelet scheme for parabolic and hyperbolic PDE's. Nonlinear Anal. **63**, e345–e356 (2005)
17. B.V. Rathish Kumar, M. Mehra, A time accurate pseudo-wavelet scheme fr two-dimensional turbulence. Int. J. Wavelets, Multiresolut. Inf. Process. **3**(4), 587–599 (2005)
18. U. Lepik, Numerical solution of evolution equations by the Haar wavelet method. Appl. Math. Comput. **185**, 695–704 (2007)

Chapter 3
Shifted Chebyshev Wavelets and Shifted Legendre Wavelets—Preliminaries

Wavelet analysis, as a relatively new and emerging area in applied mathematical research, has received considerable attention in dealing with partial differential equations and fractional partial differential equations (FPDEs) [1–9]. This chapter provides a brief introduction to shifted Chebyshev wavelets and shifted Legendre wavelets.

3.1 Introduction to Shifted Second Kind Chebyshev Wavelet Method (S2KCWM)

3.1.1 Some Properties of Second Kind Chebyshev Polynomials and Their Shifted Forms

It is well known that the second kind Chebyshev polynomials are defined on $[-1, 1]$ by

$$U_n(x) = \frac{\sin(n+1)\theta}{\sin\theta}, \quad x = \cos\theta. \tag{3.1}$$

These polynomials are orthogonal on $[-1, 1]$

$$\int_{-1}^{1} \sqrt{1-x^2}\, U_m(x) U_n(x)\mathrm{d}x = \begin{cases} 0, & m \neq n, \\ \frac{\pi}{2}, & m = n. \end{cases} \tag{3.2}$$

© Springer Nature Singapore Pte Ltd. 2019
G. Hariharan, *Wavelet Solutions for Reaction–Diffusion Problems in Science and Engineering*, Forum for Interdisciplinary Mathematics,
https://doi.org/10.1007/978-981-32-9960-3_3

The following properties of second kind Chebyshev polynomials are of fundamental importance in the sequel. They are eigenfunctions of the following singular
Sturm–Liouville equation.

$$\left(1 - x^2\right)D^2\phi_k(x) - 3xD\phi_k(x) + k(k+2)\phi_k(x) = 0, \tag{3.3}$$

where $D \equiv \frac{d}{dx}$ and may be generated by using the recurrence relation

$$U_{k+1}(x) = 2xU_k(x) - U_{k-1}(x), \quad k = 1, 2, 3, \ldots \tag{3.4}$$

Here $U_0(x) = 1$, $U_1(x) = 2x$, or Rodrigues' formula for $U_n(x)$ is given by

$$U_n(x) = \frac{(-2)^n(n+1)!}{(2n+1)!\sqrt{(1-x^2)}}D^n\left[\left(1 - x^2\right)^{n+\frac{1}{2}}\right]. \tag{3.5}$$

Theorem 3.1 *The first derivative of second kind Chebyshev polynomials is of the
form*

$$DU_n(x) = 2 \sum_{\substack{k=0 \\ (k+n)\text{odd}}}^{n-1} (k+1)U_k(x). \tag{3.6}$$

Definition 3.1 The shifted second kind Chebyshev polynomials are defined in [0,
1] by $U_n^*(x) = U_n(2x - 1)$. All results of second kind Chebyshev polynomials can
be quickly transformed to give the corresponding results for their shifted forms. The
orthogonal relation concerning the weight function $\sqrt{x - x^2}$ is given by

$$\int_0^1 \sqrt{x - x^2}\,U_n^*(x)U_m^*(x)dx = \begin{cases} 0, & m \neq n, \\ \frac{\pi}{8}, & m = n. \end{cases} \tag{3.7}$$

Corollary *The first derivative of the shifted second kind Chebyshev polynomial is
given by*

$$DU_n^*(x) = 4 \sum_{\substack{k=0 \\ (k+n)\text{odd}}} (k+1)U_k^*(x). \tag{3.8}$$

3.1.2 Shifted Second Kind Chebyshev Wavelets

These wavelets are constructed using shifted second kind Chebyshev polynomials. They are denoted by $\psi_{nm}(t) = \psi(k, n, m, t)$ and have four arguments: K, n—any positive integer, m—order of second kind Chebyshev polynomials, and t—normalized time. They are defined on the interval $[0, 1]$ by

$$\psi_{nm}(t) = \begin{cases} \frac{2^{\left(\frac{k}{2}+1\right)}}{\sqrt{\pi}} U_m^* \left(2^{k+1}t - 2n - 1\right), & t \in \left[\frac{n}{2^k}, \frac{n+1}{2^k}\right] \\ 0, & \text{Otherwise,} \end{cases} \tag{3.9}$$

$$m = 0, \ldots, M, \quad n = 0, 1, \ldots, 2^k - 1. \tag{3.10}$$

3.2 Function Approximation

A function $f(t)$ defined over $[0, 1]$ may be expanded in terms of second kind Chebyshev wavelets as

$$f(t) = \sum_{n=0}^{\infty} \sum_{m=0}^{\infty} c_{nm} \psi_{nm}(t), \tag{3.11}$$

where

$$c_{nm} = (f(t), \psi_{nm}(t))_w = \int_0^1 \sqrt{t - t^2} f(t) \psi_{nm}(t) dt. \tag{3.12}$$

If the infinite series is truncated, then it can be written as

$$f(t) = \sum_{n=0}^{2^k-1} \sum_{m=0}^{M} c_{nm} \psi_{nm}(t) = C^{\mathrm{T}} \psi(t), \tag{3.13}$$

where C and $\psi(t)$ are $2^k(M + 1) \times 1$ defined by

$$\left. \begin{aligned} C &= \left[c_{0,0}, c_{0,1}, \ldots c_{0,M}, \ldots, c_{2^k-1,M}, \ldots c_{2^k-1,1}, \ldots, c_{2^k-1,M} \right]^{\mathrm{T}} \\ \psi(t) &= \left[\psi_{0,0}, \psi_{0,1}, \ldots, \psi_{0,M}, \ldots \psi_{2^k-1,M}, \ldots, \psi_{2^k-1,1}, \ldots, \psi_{2^k-1,M} \right]^{\mathrm{T}} \end{aligned} \right\} \tag{3.14}$$

A shifted second kind Chebyshev wavelets' operational matrix of the first derivative is stated and proved in the following theorem.

Theorem 3.2 *Let $\Psi(t)$ be the second kind Chebyshev wavelet vector defined in Eq. (3.14). Then the first derivative of the vector $\Psi(t)$ can be expressed as*

$$\frac{d\psi(t)}{dt} = D\,\psi(t), \tag{3.15}$$

where D is $2^k(M + 1)$ square matrix of derivatives and is defined by

$$D = \begin{bmatrix} F_1 & O & \cdots & O \\ O & F_2 & \cdots & O \\ \vdots & \vdots & \cdots & \vdots \\ O & O & \cdots & F_n \end{bmatrix}$$

in which F is an $(M + 1)$ square matrix and its (r, s)th element is defined by

$$F_{r,s} = \begin{cases} 2^{k+2}s, & r \geq 2, r > s \text{ and } (r+s) \text{ is odd.} \\ 0, & \text{otherwise} \end{cases} \tag{3.16}$$

Corollary *The operational matrix for the nth derivative can be obtained from*

$$\frac{d^n \psi(t)}{dt^n} = D^n \psi(t), \quad n = 1, 2, \ldots \text{ where } D^n \text{ is the nth power of } D. \tag{3.17}$$

3.2.1   Operational Matrices of Derivatives for M = 2, k = 0

From the above theorem and f corollary, we have

$$D = \begin{bmatrix} 0 & 0 & 0 \\ 4 & 0 & 0 \\ 0 & 8 & 0 \end{bmatrix}, \quad D^2 = \begin{bmatrix} 0 & 0 & 0 \\ 0 & 0 & 0 \\ 32 & 0 & 0 \end{bmatrix},$$

$$\psi(t) = \sqrt{\frac{2}{\pi}} \begin{bmatrix} 2 \\ 8t - 4 \\ 32t^2 - 32t + 6 \end{bmatrix}, \quad C^T = \sqrt{\frac{\pi}{2}} [c_0 \quad c_1 \quad c_2].$$

3.2.2   *Operational Matrices of Derivatives for* k = 0, M = 3

$$
D = \begin{bmatrix} 0 & 0 & 0 & 0 \\ 4 & 0 & 0 & 0 \\ 0 & 8 & 0 & 0 \\ 4 & 0 & 12 & 0 \end{bmatrix}, \quad D^2 = \begin{bmatrix} 0 & 0 & 0 & 0 \\ 0 & 0 & 0 & 0 \\ 32 & 0 & 0 & 0 \\ 0 & 96 & 0 & 0 \end{bmatrix}, \quad D^3 = \begin{bmatrix} 0 & 0 & 0 & 0 \\ 0 & 0 & 0 & 0 \\ 0 & 0 & 0 & 0 \\ 384 & 0 & 0 & 0 \end{bmatrix},
$$

$$
\psi(t) = \sqrt{\frac{2}{\pi}} \begin{bmatrix} 2 \\ 8t - 4 \\ 32t^2 - 32t + 6 \\ 128t^3 - 192t^2 + 80t - 8 \end{bmatrix}, \quad C^T = \sqrt{\frac{\pi}{2}} [c_0 \quad c_1 \quad c_2 \quad c_3].
$$

$$(3.18)$$

3.3   Convergence Theorem for Chebyshev Wavelets

A function $f(t) \in L_w^2[0, 1]$ with bounded second derivative, say $|f''(t)| \leq L$, and let $\sum_{n=1}^{\infty} \sum_{m=0}^{\infty} c_{nm} \psi_{nm}(t)$ be its infinite second kind Chebyshev wavelet expansion, then

$$
|c_{nm}| \leq \frac{L\sqrt{\pi}}{2^3 (n+1)^{\frac{5}{2}}(m^2 + 2m - 3)}, \tag{3.19}
$$

This means the second kind Chebyshev wavelet series converges uniformly to $f(t)$ and

$$
f(t) = \sum_{n=1}^{\infty} \sum_{m=0}^{\infty} c_{nm} \psi_{nm}(t).
$$

Proof From the definition of coefficient c_{nm}, we have $c_{nm} = \int_0^1 f(t)\psi_{nm}(t)w_{nk}(t)dt$, where $w_{nk}(t) = \sqrt{t - t^2}$, which is a weight function.

$$
c_{nm} = \frac{2^{\frac{k}{2}+1}}{\sqrt{\pi}} \int_{\frac{n}{2^k}}^{\frac{n+1}{2^k}} f(t)U_m(2^{k+1}t - 2n - 1)w_{nm}(t)dt.
$$

By substituting $2^{k+1}t - 2n - 1 = \cos\theta$, we get

$$c_{nm} = \frac{2^{-\frac{k}{2}}}{\sqrt{\pi}} \int_0^\pi f\left(\frac{\cos\theta + 2n + 1}{2^{k+1}}\right) \sin(m+1)\theta \sin\theta d\theta.$$

Using integration by parts two times, we get

$$c_{nm} = \frac{1}{2^{\frac{5k}{2}+3}\sqrt{\pi}} \int_0^\pi f''\left(\frac{\cos\theta + 2n + 1}{2^{k+1}}\right) h_m(\theta) d\theta,$$

where

$$h_m(\theta) = \frac{\sin\theta}{m}\left(\frac{\sin((m-1)\theta)}{m-1} - \frac{\sin((m+1)\theta)}{m+1}\right)$$
$$- \frac{\sin\theta}{m+2}\left(\frac{\sin((m+1)\theta)}{m+1} - \frac{\sin((m+3)\theta)}{m+3}\right).$$

So we have

$$|c_{nm}| \le \frac{L2^{\left(-\frac{5k}{2}-3\right)}}{\sqrt{\pi}} \int_0^\pi |h_m(\theta)| d\theta \le \frac{L2^{\left(-\frac{5k}{2}-3\right)}\sqrt{\pi}}{(m^2 + 2m - 3)},$$

Since $n \le 2^k - 1$, we obtain

$$|c_{nm}| \le \frac{L\sqrt{\pi}}{2^{\frac{5k}{2}+3}(m^2 + 2m - 3)}$$
$$\le \frac{L\sqrt{\pi}}{2^3(n+1)^{\frac{5}{2}}(m^2 + 2m - 3)}.$$

Hence, the series $\sum_{n=1}^\infty \sum_{m=0}^\infty c_{nm}$ is convergent; it follows that $\sum_{n=1}^\infty \sum_{m=0}^\infty c_{nm}\psi_{nm}(t)$ converges to the function $f(t)$ uniformly.

3.3.1 Accuracy Estimation

Suppose $f(t)$ be a continuous function defined on $[0, 1]$, with second derivative of $f(t)$ bounded by L, then we have the following accuracy estimation

$$\|e_{M,K}(t)\|_2 \le \left(\frac{\pi L^2}{2^6}\sum_{n=0}^\infty \sum_{m=M}^\infty \frac{1}{(n+1)^5(m^2+2m-3)^2} + \frac{\pi L^2}{2^6}\sum_{n=2}^\infty \sum_{m=0}^{M-1} \frac{1}{(n+1)^5(m^2+2m-3)^2}\right)^{\frac{1}{2}},$$

$$(3.20)$$

where

$$\left\| e_{M,K}(t) \right\|_2 = \left(\int_0^1 \left(f(t) - \sum_{n=0}^{2^k-1} \sum_{m=0}^{M-1} c_{nm} \psi_{nm}(t) \right)^2 w_n(t) dt \right)^{\frac{1}{2}}. \tag{3.21}$$

Proof

We have

$$\left\| e_{M,K}(t) \right\|_2 = \int_0^1 \left(f(t) - \sum_{n=0}^{2^k-1} \sum_{m=0}^{M-1} c_{nm} \psi_{nm}(t) \right)^2 w(t) dt$$

$$= \int_0^1 \left(\sum_{n=0}^{\infty} \sum_{m=0}^{\infty} c_{nm} \psi_{nm}(t) - \sum_{n=0}^{2^k-1} \sum_{m=0}^{M-1} c_{nm} \psi_{nm}(t) \right)^2 w(t) dt$$

$$= \sum_{n=0}^{\infty} \sum_{m=M}^{\infty} c_{nm}^2 \int_0^1 \psi_{nm}^2(t) w(t) dt + \sum_{n=2^k}^{\infty} \sum_{m=0}^{M-1} c_{nm}^2 \int_0^1 \psi_{nm}^2(t) w(t) dt$$

$$= \sum_{n=0}^{\infty} \sum_{m=M}^{\infty} c_{nm}^2 + \sum_{n=2^k}^{\infty} \sum_{m=0}^{M-1} c_{nm}^2.$$

3.4 Legendre Wavelet Method (LWM)

These wavelets are constructed using Legendre polynomials. They are denoted by $\psi_{n,m}(t) = \psi(k, n, m, t)$ and have four arguments. k, n—any positive integer, m—order of Legendre polynomials, and t—normalized time. They are defined on the interval [0, 1] by

$$\psi_{nm}(t) = \begin{cases} \sqrt{m + \frac{1}{2}} 2^{\frac{k}{2}} L_m \left(2^k t - \hat{n} \right), & \text{for} \quad \frac{\hat{n}-1}{2^k} \leq t \leq \frac{\hat{n}+1}{2^k}, \\ 0, & \text{otherwise} \end{cases} \tag{3.22}$$

where $m = 0, 1, 2, \ldots, M-1$ and $k = 1, 2, \ldots, 2^k - 1$. The coefficient $\sqrt{m + \frac{1}{2}}$ is for orthonormality. Then the wavelets $\Psi_{k,m}(x)$ form an orthonormal basis for $L^2[0, 1]$. In the above formulation of Legendre wavelets, the Legendre polynomials are in the following way:

$$p_0 = 1,$$
$$p_1 = x,$$
$$p_{m+1}(x) = \frac{2m+1}{m+1} \times p_m(x) - \frac{m}{m+1} p_{m-1}(x).$$

(3.23)

Here $\{p_{m+1}(x)\}$ are the orthogonal functions of order m, which is named the well-known shifted Legendre polynomials on the interval $[0, 1]$.

Theorem 3.3 *Let $\Psi(t)$ be the Legendre wavelet vector defined in Eq. (3.22). Then the first derivative of the vector $\Psi(t)$ can be expressed as*

$$\frac{d\psi(t)}{dt} = D\,\psi(t),$$

(3.24)

where D is $2^k(M+1)$ square matrix of derivatives and is defined by

$$D = \begin{bmatrix} F_1 & O & \cdots & O \\ O & F_2 & \cdots & O \\ \vdots & \vdots & \cdots & \vdots \\ O & O & \cdots & F_n \end{bmatrix}$$

in which F is an $(M+1)$ square matrix and its (r, s)th element is defined by

$$F_{r,s} = \begin{cases} 2^k & \sqrt{(2r-1)(2s-1)}, \; r \geq 2, \; r > s, \; r+s \text{ is odd} \\ 0, & \text{Otherwise} \end{cases}$$

(3.25)

Corollary *The operational matrix for the nth derivative can be obtained from*

$$\frac{d^n \psi(t)}{dt^n} = D^n \psi(t), \quad n = 1, 2, \ldots \text{ where } D^n \text{ is the } n\text{th power of } D.$$

(3.26)

3.4.1 Operational Matrices of Derivatives for M = 2, k = 0

From the above theorem and corollary, we have

$$D = \begin{bmatrix} 0 & 0 & 0 \\ 2\sqrt{3} & 0 & 0 \\ 0 & 2\sqrt{3}\sqrt{5} & 0 \end{bmatrix}, \quad D^2 = \begin{bmatrix} 0 & 0 & 0 \\ 0 & 0 & 0 \\ 12\sqrt{5} & 0 & 0 \end{bmatrix},$$

$$\psi(t) = \sqrt{\frac{2}{\pi}} \begin{bmatrix} 1 \\ 2\sqrt{3}t - \sqrt{3} \\ 6\sqrt{5}t^2 - 6\sqrt{5}t + \sqrt{5} \end{bmatrix}, \quad C^T = \begin{bmatrix} c_0 & c_1 & c_2 \end{bmatrix}. \tag{3.27}$$

For $M = 3$, $k = 0$, the operational matrices of derivatives are

$$D = \begin{bmatrix} 0 & 0 & 0 & 0 \\ 2\sqrt{3} & 0 & 0 & 0 \\ 0 & 2\sqrt{3}\sqrt{5} & 0 & 0 \\ 2 & 0 & 2\sqrt{5}\sqrt{7} & 0 \end{bmatrix}, \quad D^2 = \begin{bmatrix} 0 & 0 & 0 & 0 \\ 0 & 0 & 0 & 0 \\ 12\sqrt{5} & 0 & 0 & 0 \\ 0 & 20\sqrt{3}\sqrt{7} & 0 & 0 \end{bmatrix},$$

$$D^3 = \begin{bmatrix} 0 & 0 & 0 & 0 \\ 0 & 0 & 0 & 0 \\ 0 & 0 & 0 & 0 \\ 120\sqrt{7} & 0 & 0 & 0 \end{bmatrix},$$

$$\psi(t) = \sqrt{\frac{2}{\pi}} \begin{bmatrix} 1 \\ 2\sqrt{3}t - \sqrt{3} \\ 6\sqrt{5}t^2 - 6\sqrt{5}t + \sqrt{5} \\ 20\sqrt{7}x^3 - 30\sqrt{7}x^2 + 12\sqrt{7}x - \sqrt{7} \end{bmatrix}, \quad C^T = \begin{bmatrix} c_0 & c_1 & c_2 c_3 \end{bmatrix}. \tag{3.28}$$

3.5 Convergence Theorem for Legendre Wavelets

If a continuous function $f(t)$ defined in $[0, 1]$ has $|f''(t)| \leq M$ can be expressed as an infinite sum of Legendre wavelets and the series converges uniformly to $f(t)$, that is $f(t) = \sum_{n=1}^{\infty} \sum_{m=0}^{\infty} c_{nm}\psi_{nm}(t)$, where $c_{nm} = \langle f(t), \psi_{nm}(t) \rangle$ then

$$|c_{nm}| < \frac{\sqrt{12}M}{(2n)^{\frac{5}{2}}(2m-3)^2}, \quad m > 1, \ n \geq 0. \tag{3.29}$$

Proof $c_{nm} = \int\limits_0^1 f(t)\psi_{nm}(t)dt = \int\limits_{\frac{\hat{n}-1}{2^k}}^{\frac{\hat{n}+1}{2^k}} f(t)\left(\frac{2m+1}{2}\right)^{\frac{1}{2}}2^{\frac{k}{2}}p_m\left(2^k t - \hat{n}\right)dt.$

Now, let $2^k t - \hat{n} = u$, then $dt = \frac{1}{2^k}du$

$$c_{nm} = \int\limits_{-1}^1 f\left(\frac{\hat{n}+u}{2^k}\right)\left(\frac{2m+1}{2}\right)^{\frac{1}{2}}2^{\frac{k}{2}}p_m(u)\frac{1}{2^k}du$$

$$= \left(\frac{2m+1}{2^{K+1}}\right)^{\frac{1}{2}}\int\limits_{-1}^1 f\left(\frac{\hat{n}+u}{2^k}\right)p_m(u)du$$

$$= \left(\frac{1}{2^{k+1}(2m+1)}\right)^{\frac{1}{2}}\int\limits_{-1}^1 f\left(\frac{\hat{n}+u}{2^k}\right)d(p_{m+1}(u) - p_{m-1}(u))$$

$$= \left(\frac{1}{2^{k+1}(2m+1)}\right)^{\frac{1}{2}}\left(f\left(\frac{\hat{n}+u}{2^k}\right)(p_{m+1}(u) - p_{m-1}(u))\right)\Big|_{-1}^1$$

$$- \left(\frac{1}{2^{k+1}(2m+1)}\right)^{\frac{1}{2}}\int\limits_{-1}^1 f'\left(\frac{\hat{n}+u}{2^k}\right)\frac{1}{2^k}(p_{m+1}(u) - p_{m-1}(u))du$$

$$= -\left(\frac{1}{2^{k+1}(2m+1)}\right)^{\frac{1}{2}}\int\limits_{-1}^1 f'\left(\frac{\hat{n}+u}{2^k}\right)\frac{1}{2^k}(p_{m+1}(u) - p_{m-1}(u))du$$

$$= -\left(\frac{1}{2^{3k+1}(2m+1)}\right)^{\frac{1}{2}}\int\limits_{-1}^1 f'\left(\frac{\hat{n}+u}{2^k}\right)(p_{m+1}(u) - p_{m-1}(u))du$$

$$= -\left(\frac{1}{2^{3k+1}(2m+1)}\right)^{\frac{1}{2}}\int\limits_{-1}^1 f'\left(\frac{\hat{n}+u}{2^k}\right)d\left(\frac{p_{m+2}(u) - p_m(u)}{2m+3} - \frac{p_m(u) - p_{m-2}(u)}{2m-1}\right)$$

$$= -\left(\frac{1}{2^{3k+1}(2m+1)}\right)^{\frac{1}{2}}\left(f'\left(\frac{\hat{n}+u}{2^k}\right)\left(\frac{p_{m+2}(u) - p_m(u)}{2m+3} - \frac{p_m(u) - p_{m-2}(u)}{2m-1}\right)\right)\Big|_{-1}^1$$

$$+ \left(\frac{1}{2^{5k+1}(2m+1)}\right)^{\frac{1}{2}}\int\limits_{-1}^1 f''\left(\frac{\hat{n}+u}{2^k}\right)\left(\frac{p_{m+2}(u) - p_m(u)}{2m+3} - \frac{p_m(u) - p_{m-2}(u)}{2m-1}\right)du$$

$$= \left(\frac{1}{2^{5k+1}(2m+1)}\right)^{\frac{1}{2}}\int\limits_{-1}^1 f''\left(\frac{\hat{n}+u}{2^k}\right)\left(\frac{p_{m+2}(u) - p_m(u)}{2m+3} - \frac{p_m(u) - p_{m-2}(u)}{2m-1}\right)du.$$

Consider

$$\left| \int_{-1}^{1} f''\left(\frac{\hat{n}+u}{2^k}\right) \left(\frac{p_{m+2}(u) - p_m(u)}{2m+3} - \frac{p_m(u) - p_{m-2}(u)}{2m-1}\right) du \right|^2$$

$$= \left| \int_{-1}^{1} f''\left(\frac{\hat{n}+u}{2^k}\right) \left(\frac{(2m-1)p_{m+2}(u) - (4m+2)p_m(u) + (2m+3)p_{m-2}}{(2m+3)(2m-1)}\right) du \right|^2$$

$$\leq \left| \int_{-1}^{1} f''\left(\frac{\hat{n}+u}{2^k}\right) \right|^2 du \int_{-1}^{1} \left| \left(\frac{(2m-1)p_{m+2}(u) - (4m+2)p_m(u) + (2m+3)p_{m-2}}{(2m+3)(2m-1)}\right) \right|^2 du$$

$$< 2M^2 \int_{-1}^{1} \frac{(2m-1)^2 p_{m+2}^2(u) + (4m+2)^2 p_m^2(u) + (2m+3)^2 p_{m-2}^2(u)}{(2m+3)^2(2m-1)^2} du$$

$$= \frac{2M^2}{(2m+3)^2(2m-1^2)} \left[(2m-1)^2 \frac{2}{2m+5} + (4m+2)^2 \frac{2}{2m+1} + (2m+3)^2 \frac{2}{2m-3}\right]$$

$$< \frac{2M^2}{(2m+3)^2(2m-1)^2} \left[\frac{2(2m-1)^2 + 8(2m+1)^2 + 2(2m+3)^2}{2m-3}\right]$$

$$< \frac{2M^2}{(2m+3)^2(2m-1)^2} \frac{12(m+3)^2}{2m-3}$$

$$= \frac{24M^2}{(2m-1)^2(2m-3)}.$$

Thus, we get

$$\left| \int_{-1}^{1} f''\left(\frac{\hat{n}+u}{2^k}\right) \left(\frac{p_{m+2}(u) - p_m(u)}{2m+3} - \frac{p_m(u) - p_{m-2}(u)}{2m-1}\right) du \right| < \frac{\sqrt{24}M}{(2m-1)\sqrt{(2m-3)}}.$$

Therefore, we have

$$|c_{nm}| < \frac{\sqrt{12}M}{2^{\frac{5k}{2}}} \frac{1}{\sqrt{(2m+1)}} \frac{1}{(2m-1)\sqrt{(2m-3)}}$$

$$< \sqrt{12} \frac{1}{2^{\frac{5k}{2}}} \frac{1}{(2m-3)^2} \tag{3.30}$$

$$< \frac{\sqrt{12}}{(2n)^{\frac{5}{2}}(2m-3)^2}.$$

Hence, the series $\sum_{n=1}^{\infty} \sum_{m=0}^{\infty} c_{nm}$ is convergent; it follows that $\sum_{n=1}^{\infty} \sum_{m=0}^{\infty} c_{nm} \psi_{nm}(t)$ converges to the function $f(t)$ uniformly.

3.6 Error Analysis

Suppose $f(t)$ be a continuous function defined on $[0, 1]$ with second derivatives $f''(t)$ bounded by M; then, we have the following accuracy estimation

$$\left\|e_{M,K}(t)\right\|_2 \leq \left(\frac{3M^2}{2}\sum_{n=0}^{\infty}\sum_{m=M}^{\infty}\left(\frac{1}{n^5(2m-3)^4}\right) + \frac{3M^2}{2}\sum_{n=2}^{\infty}\sum_{m=0}^{M-1}\left(\frac{1}{n^5(2m-3)^4}\right)\right)^{\frac{1}{2}},$$

(3.31)

where

$$\left\|e_{M,K}(t)\right\|_2 = \left(\int_0^1\left(f(t) - \sum_{n=0}^{2^k-1}\sum_{m=0}^{M-1}c_{nm}\psi_{nm}(t)\right)^2 dt\right)^{\frac{1}{2}}.$$

(3.32)

Proof We have

$$\left\|e_{M,K}(t)\right\|_2 = \int_0^1\left(f(t) - \sum_{n=0}^{2^k-1}\sum_{m=0}^{M-1}c_{nm}\psi_{nm}(t)\right)^2 dt$$

$$= \int_0^1\left(\sum_{n=0}^{\infty}\sum_{m=0}^{\infty}c_{nm}\psi_{nm}(t) - \sum_{n=0}^{2^k-1}\sum_{m=0}^{M-1}c_{nm}\psi_{nm}(t)\right)^2 dt$$

$$= \sum_{n=0}^{\infty}\sum_{m=M}^{\infty}c_{nm}^2\int_0^1\psi_{nm}^2(t)dt + \sum_{n=2^k}^{\infty}\sum_{m=0}^{M-1}c_{nm}^2\int_0^1\psi_{nm}^2(t)dt$$

$$= \sum_{n=0}^{\infty}\sum_{m=M}^{\infty}c_{nm}^2 + \sum_{n=2^k}^{\infty}\sum_{m=0}^{M-1}c_{nm}^2.$$

3.7 2-D Legendre Wavelets

2-D Legendre wavelets in $L^2(R)$ over the interval $[0, 1] \times [0, 1]$ as the form

$$\Psi_{n,m,n',m'}(x, y) = \begin{cases} \sqrt{\left(m+\frac{1}{2}\right)\left(m'+\frac{1}{2}\right)}2^{\frac{k+k'}{2}}p_m(x)p_{m'}(y), \\ \frac{n-1}{2^{k-1}} \leq x \leq \frac{n}{2^{k-1}}, \frac{n'-1}{2^{k'-1}} \leq y \leq \frac{n'}{2^{k'-1}}; \\ 0, \quad \text{otherwise.} \end{cases}$$

(3.33)

and $m = 0, 1, 2, \ldots, M-1$, $m' = 0, 1, 2, 3, \ldots, M'-1$, $n = 1, 2, \ldots, 2^{k-1}$, $n' = 1, 2,$ $\ldots, 2^{k'-1}$,

where

$$P_m(x) = \overline{P_{m'}}\left(2^k x - 2n + 1\right), P_{m'}(y) = \overline{P_{m'}}\left(2^{k'} y - 2n' + 1\right), \tag{3.34}$$

$\overline{P_m}$ are Legendre functions of order m defined over the interval $[-1, 1]$.

With the help of 2-D shifted Legendre polynomials into $x \in \left[\frac{n-1}{2^{k-1}}, \frac{n}{2^{k-1}}\right]$, and $y \in \left[\frac{n'-1}{2^{k'-1}}, \frac{n'}{2^{k'-1}}\right]$, the $\int_0^1 \Psi_{n,m,n',m'}(x, y)$ can be written as

$$\int\limits_0^1 \Psi_{n,m,n',m'}(x, y) = A_{m,m'} \cdot P_{m'}(x) P_{m'}(y) \chi_{\left[\frac{n-1}{2^{k-1}}, \frac{n}{2^{k-1}}\right]\atop \left[\frac{n'-1}{2^{k'-1}}, \frac{n'}{2^{k'-1}}\right]}(x, y), \tag{3.35}$$

in which $A_{m,m'} = \sqrt{\left(m + \frac{1}{2}\right)\left(m' + \frac{1}{2}\right)} 2^{\frac{k+k'}{2}}$ and $\chi_{\left[\frac{n-1}{2^{k-1}}, \frac{n}{2^{k-1}}\right]\atop \left[\frac{n'-1}{2^{k'-1}}, \frac{n'}{2^{k'-1}}\right]}(x, y)$ is a characteristic

function defined as $\chi_{\left[\frac{n-1}{2^{k-1}}, \frac{n}{2^{k-1}}\right]\atop \left[\frac{n'-1}{2^{k'-1}}, \frac{n'}{2^{k'-1}}\right]}(x, y) = \begin{cases} 1, x \in \left[\frac{n-1}{2^{k-1}}, \frac{n}{2^{k-1}}\right], y \in \left[\frac{n'-1}{2^{k'-1}}, \frac{n'}{2^{k'-1}}\right]; \\ 0, \text{ otherwise} \end{cases}$

Two-dimensional Legendre wavelets are an orthonormal set over $[0, 1] \times [0, 1]$:

$$\int\limits_0^1 \int\limits_0^1 \Psi_{n,m,n',m'}(x, y)\, \Psi_{n_1,m_1,n'_1,m'_1}(x, y)\, dxdy = \delta_{n,n_1} \delta_{n',n'_1} \delta_{m',m'_1}. \tag{3.36}$$

The function $u(x, y) \in L^2(R)$ defined over $[0, 1] \times [0, 1]$ may be expanded as

$$u(x, y) = X(x)Y(y) \cong \sum_{n=1}^{\infty}\sum_{m=0}^{\infty}\sum_{n'=1}^{\infty}\sum_{m'=0}^{\infty} c_{n,m,n',m'} \Psi_{n,m,n',m'}(x, y). \tag{3.37}$$

If the infinite series in Eq. (3.37) is truncated, then Eq. (3.37) can be written as

$$u(x, y) = X(x)Y(y) \cong \sum_{n=1}^{2^{k-1}}\sum_{m=0}^{M-1}\sum_{n'=1}^{2^{k'-1}}\sum_{m'=0}^{M'-1} c_{n,m,n',m'} \Psi_{n,m,n',m'}(x, y), \tag{3.38}$$

where $c_{n,m,n',m'} = \int\limits_0^1 \int\limits_0^1 X(x)Y(y)\Psi_{n,m,n',m'}(x, y)dxdy.$

Equation (3.38) can be expressed as

$$u(x,\ y) = c^{\mathrm{T}} \cdot \Psi(x,\ y),\tag{3.39}$$

where C and $\Psi(x, y)$ are coefficient matrix and wavelet vector matrix, respectively. The number of dimensions of C and $\Psi(x, y)$ is $2^{k-1}2^{k'-1}MM' \times 1$ and is given by

$$
\begin{aligned}
C = \Big[& c_{1,0,1,0},\ \cdots,\ c_{1,0,1,M'-1}, c_{1,0,2,0},\ \cdots,\ c_{1,0,2,M'-1},\ \cdots,\ c_{1,0,2^{k'-1},0},\ \cdots, \\
& c_{1,0,2^{k'-1},M'-1},\ \cdots,\ c_{1,M-1,1,0},\ \cdots,\ c_{1,M-1,1,M'-1}, c_{1,M-1,2,0},\ \cdots, \\
& c_{1,M-1,2,M'-1},\ \cdots,\ c_{1,M-1,2^{k-1},0},\ \cdots,\ c_{1,M-1,2^{k-1},M'-1},\ \cdots, c_{2,0,1,0},\ \cdots, \\
& c_{2,0,1,M'-1}, c_{2,0,2,0},\ \cdots,\ c_{2,0,2,M'-1},\ \cdots,\ c_{2,0,2^{k-1},0},\ \cdots,\ c_{2,0,2^{k-1},M'-1},\ \cdots, \\
& c_{2,M-1,1,0},\ \cdots,\ c_{2,M-1,1,M'-1}, c_{2,M-1,2,0},\ \cdots,\ c_{2,M-1,2,M'-1},\ \cdots, \\
& c_{2,M-1,2^{k-1},0},\ \cdots,\ c_{2,M-1,2^{k-1},M'-1},\ \cdots,\ c_{2^{k-1},0,1,0},\ \cdots, c_{2^{k-1},0,1M'-1}, \\
& c_{2^{k-1},0,2,0},\ \cdots,\ c_{2^{k-1},0,,M'-1},\ \cdots,\ c_{2^{k-1},0,2^{k-1},0},\ \cdots,\ c_{2^{k-1},M-1,2^{k'-1},M'-1} \Big]^{\mathrm{T}}
\end{aligned}\tag{3.40}
$$

$$
\begin{aligned}
\Psi = \Big[& \Psi_{1,0,1,0},\ \cdots,\ \Psi_{1,0,1,M'-1}, \Psi_{1,0,2,0},\ \cdots,\ \Psi_{1,0,2^{k-1},0},\ \cdots, \\
& \Psi_{1,0,2^{k'-1},M'-1},\ \cdots,\ \Psi_{1,M-1,1,0},\ \cdots,\ \Psi_{1,M-1,1,M'-1}, \\
& \Psi_{1,M-1,2,0},\ \cdots,\ \Psi_{1,M-1,2,M'-1},\ \cdots,\ \Psi_{1,M-1,2^{k-1},0},\ \cdots, \\
& \Psi_{1,M-1,2^{k-1},M'-1},\ \cdots,\ \Psi_{2,0,1,0},\ \cdots,\ \Psi_{2,0,1,M'-1}, \Psi_{2,0,2,0},\ \cdots, \\
& \Psi_{2,0,2,M'-1},\ \cdots,\ \Psi_{2,0,2^{k'-1},0},\ \cdots,\ \Psi_{2,0,2^{k-1},M'-1},\ \cdots, \\
& \Psi_{2,M-1,1,0},\ \cdots,\ \Psi_{2,M-1,1,M'-1}, \Psi_{2,M-1,2,0},\ \cdots, \\
& \Psi_{2,M-1,2,M'-1},\ \cdots,\ \Psi_{2,M-1,2^{k'-1},0},\ \cdots,\ \Psi_{2,M-1,2^{k'-1},M'-1}, \\
& \Psi_{2^{k-1},0,1,0},\ \cdots,\ \Psi_{2^{k-1},0,1,M'-1}, \Psi_{2^{k-1},0,2,0},\ \cdots, \\
& \Psi_{2^{k-1},0,2,M'-1},\ \cdots,\ \Psi_{2^{k-1},0,2^{k-1},0},\ \cdots,\ \Psi_{2^{k-1},M-1,2^{k-1},M'-1} \Big]^{\mathrm{T}}
\end{aligned}\tag{3.41}
$$

The integration of the product of two Legendre wavelet function vectors is obtained as

$$\int\limits_{0}^{1}\int\limits_{0}^{1} \Psi(x,y)\Psi^{\mathrm{T}}(x,\ y)\mathrm{d}x\mathrm{d}y = I,\tag{3.42}$$

where I is an identity matrix.

A 2-D function $f(x, y)$ defined $[0, 1) \times [0, 1)$ may be expanded by Legendre wavelet series as

$$f(x,y) = \sum_{i=1}^{2^{k}M}\sum_{j=1}^{2^{k}M} C_{ij}\Psi_i(x)\Psi_j(y) = \Psi^{\mathrm{T}}(x)C\Psi(y),\tag{3.43}$$

where

$$C_{ij} = \int_0^1 f(x,y)\Psi_i(x)dx \int_0^1 f(x,y)\Psi_j(y)dt. \tag{3.44}$$

Equation (3.43) becomes

$$f(x,y) = \Psi^T(x)C\Psi(y), \tag{3.45}$$

where C and $\Psi(t)$ are $2^{k-1}M \times 1$ matrices given by

$$C = \begin{bmatrix} c_{0,0} & c_{0,1} & \cdots & c_{0,2^{k-1}M} \\ c_{1,0} & c_{1,1} & \cdots & c_{1,2^{k-1}M} \\ \vdots & \vdots & \ddots & \vdots \\ c_{2^{k-1}M,0} & c_{2^{k-1}M,1} & \cdots & c_{2^{k-1}M2^{k-1}M} \end{bmatrix}$$

Theorem 3.4 Let $\Psi(x,y)$ be the 2-D Legendre wavelet vector defined in Eq. (3.33). Then

$$\frac{\partial \Psi(x,y)}{\partial x} = D_x \Psi(x,y). \tag{3.46}$$

where D_x is $2^{k-1} 2^{k'-1}MM' \times 2^{k-1}2^{k'-1}MM'$.

$$D_x = \begin{bmatrix} D & O' & \cdots & O' \\ O' & D & \cdots & O' \\ \vdots & \vdots & \ddots & \vdots \\ O' & O' & \cdots & D \end{bmatrix} \text{ in which '}O\text{' and } D \text{ are } 2^{k-1}M2^{k-1}MM \times 2^{k-1}M2^{k-1}MM'$$

matrices, and the element of D is defined as follows:

$$D_{r,s} = \begin{cases} 2^k\sqrt{(2r-1)(2s-1)}I, & r = 2,\ 3,\ \ldots,\ M;\ s = 1,\ \ldots,\ r-1;\ r+s \text{ is odd}, \\ 0 & \text{otherwise.} \end{cases} \tag{3.47}$$

and I, O are $2^{k'-1}M' \times 2^{k'-1}M'$ identity matrices.

Theorem 3.5 Let $\Psi(x,y)$ be the 2-D Legendre wavelet vector defined in Eq. (3.33), we have

$$\frac{\partial \Psi(x,y)}{\partial x} = D_y \Psi(x,y), \tag{3.48}$$

$$D_y = \begin{bmatrix} D & O' & \cdots & O' \\ O' & D & \cdots & O' \\ \vdots & \vdots & \ddots & \vdots \\ O' & O' & \cdots & D \end{bmatrix},$$

where D_y is $2^{k-1} 2^{k'-1} MM'$ x $2^{k-1} 2^{k'-1} MM'$ in which D is $MM' \times MM'$ matrix given as

$$D = \begin{bmatrix} F & O & \cdots & O \\ O & F & \cdots & O \\ \vdots & \vdots & \ddots & \vdots \\ O & O & \cdots & F \end{bmatrix},$$

in which O and F are $M' \times M'$ matrix and F is defined as follows:

$$F_{r,s} = \begin{cases} 2^{k'} \sqrt{(2r-1)(2s-1)}, & r = 2, \ldots, M'; \ S = 1, \ldots, r-1; \text{ and } r+s \text{ is odd,} \\ 0 & \text{otherwise} \end{cases}$$

$$(3.49)$$

3.8   Block-Pulse Functions (BPFs)

Block-pulse function (BPF) forms a complete set of orthogonal functions which are defined on the interval $[0, b)$ by

$$b_i(t) \begin{cases} 1, & \frac{i-1}{m} b \leq t < \frac{i}{m} b \\ 0, & \text{otherwise} \end{cases} \tag{3.50}$$

for $i = 1, 2, \ldots, m$. It is also known that for any absolutely integrable function $f(t)$ on $[0, b)$ can be expanded in block-pulse functions:

$$f(t) \cong \xi^T B_m(t), \tag{3.51}$$

$$\xi^T = [f_1, f_2, \ldots, f_m], B_m(t) = [b_1(t), b_2(t), \ldots, b_m(t)], \tag{3.52}$$

where f_i are the coefficients of the block-pulse function and are given by

$$f_i = \frac{m}{b} \int_0^b f(t) b_i(t) \mathrm{d}t. \tag{3.53}$$

Remark 1

Let A and B are two matrices of m x m; then $A \otimes B = \left(a_{ij} \times b_{ij} \right)_{mm}$.

Lemma 1 *Assuming f(t) and g(t) are two absolutely integrable functions, which can be expanded in block-pulse function as f(t) = FB(t) and g(t) = GB(t) respectively, then we have*

$$f(t)g(t) = FB(t)B^{\mathrm{T}}(t)G^{\mathrm{T}} = HB(t), \quad \text{where } H = F \otimes G. \quad (3.54)$$

3.9 Approximating the Nonlinear Term

The Legendre wavelets can be expanded into m-set of block-pulse functions as

$$\Psi(t) = \emptyset_{m \times m} B_m(t). \quad (3.55)$$

Taking the collocation points as following

$$t_i = \frac{i - 1/2}{2^{k-1}M}, \quad i = 1, 2, \ldots, 2^{k-1}M \quad (3.56)$$

The m-square Legendre matrix $\emptyset_{m \times m}$ is defined as

$$\emptyset_{m \times m} \cong \left[\Psi(t_1) \Psi(t_2) \ldots \Psi(t_{2^{k-1}M}) \right]. \quad (3.57)$$

The operational matrix of a product of Legendre wavelets can be obtained by using the properties of BPFs; let $f(x, t)$ and $g(x, t)$ are two integrable functions, which can be expanded by Legendre wavelets as $f(x, t) = \Psi^{\mathrm{T}}(x) F \Psi(t)$ and $g(x, t) = \Psi^{\mathrm{T}}(x) G \Psi(t)$, respectively.

$$f(x, t) = \Psi^{\mathrm{T}}(x) F \Psi(t) = B^{\mathrm{T}}(x) \emptyset_{mm}^{\mathrm{T}} F \emptyset_{mm} B(t), \quad (3.58)$$

$$g(x, t) = \Psi^{\mathrm{T}}(x) G \Psi(t) = B^{\mathrm{T}}(x) \emptyset_{mm}^{\mathrm{T}} G \emptyset_{mm} B(t), \quad (3.59)$$

and $F_b = \emptyset_{mm}^{\mathrm{T}} F \emptyset_{mm}, G_b = \emptyset_{mm}^{\mathrm{T}} G \emptyset_{mm}, H_b = F_b \otimes G_b$.
Then,

$$
\begin{aligned}
f(x, t)g(x, t) &= B^{\mathrm{T}} H_b B(t) \\
&= B^{\mathrm{T}}(x) \emptyset_{mm}^{\mathrm{T}} inv(\emptyset_{mm}^{\mathrm{T}}) H_b inv\left(inv(\emptyset_{mm}^{\mathrm{T}})\right) H_b inv(\emptyset_{mm}) \emptyset_{mm} B(t) \quad (3.60) \\
&= \Psi^{\mathrm{T}}(x) H \Psi(t),
\end{aligned}
$$

where $H = inv(\emptyset_{mm}^{\mathrm{T}}) H_b inv((\emptyset_{mm}))$.

3.10 Approximation of Function

A function $f(x)$ with the domain $[0, 1]$ can be approximated by:

$$f(x) = \sum_{k=1}^{\infty} \sum_{m=0}^{\infty} c_{k,m} \Psi_{k,m}(x) = C^{\mathrm{T}}.\Psi(x). \tag{3.61}$$

If the infinite series in Eq. (3.61) is truncated, then this equation can be written as:

$$f(x) \simeq \sum_{k=1}^{\infty} \sum_{m=0}^{\infty} c_{k,m} \Psi_{k,m}(x) = C^{\mathrm{T}} \cdot \Psi(x), \tag{3.62}$$

where C and Ψ are the matrices of size $(2^{j-1} M \times 1)$.

$$C = \left[c_{1,0},\ c_{1,1},\ \dots,\ c_{1,M-1},\ c_{2,0},\ c_{2,1},\ \dots,\ c_{2,M-1},\ \dots,\ c_2^{j-1},\ 1,\ \dots,\ c_2^{j-1},\ M-1 \right]^{\mathrm{T}} \tag{3.63}$$

$$\Psi(x) = \left[\Psi_{1,0},\ \Psi_{1,1},\ \Psi_{2,0},\ \Psi_{2,1},\ \dots,\ \Psi_{2,M-1},\ \dots,\ \Psi_{2^{j-1},M-1} \right]^{\mathrm{T}}. \tag{3.64}$$

References

1. S.S. Ray, The transport dynamics induced by riesz potential in modelling fractional reaction-diffusion-mechanics system. J. Comput Nonlinear Dyn. **13**(2) (2018). https://doi.org/10.1115/1.4037418
2. S. Sahoo, S.S. Ray, The conservation laws with Lie symmetry analysis for time fractional integrable coupled KdV-mKdV system. Int. J. Non-Linear Mech. **98**, 114–121 (2018). https://doi.org/10.1016/j.ijnonlinmec.2017
3. M. Kumar, S. Pandit, Wavelet transform and wavelet based numerical methods: an introduction. Int. J Nonlinear Sci. **13**(3), 325–345 (2012)
4. V. Kumar, M. Mehra, Cubic spline adaptive wavelet scheme to solve singularly perturbed reaction-diffusion problems. Int. J. Wavelets Multiresolut. Inf. Process. **5**(2), 317–331 (2007)
5. G. Hariharan, K. Kannan, K.R. Sharma, Haar wavelet method for solving Fisher's equation. Appl. Math. Comput. **211**, 284–292 (2009)
6. G. Hariharan, K. Kannan, Review of wavelet methods for the solution of reaction–diffusion problems in science and engineering. Appl. Math. Model. **38**(1), 799–813 (2014)
7. G. Hariharan, K. Kannan, An overview of Haar wavelet method for solving differential and integral equations. World Appl. Sci. J. **23**(12), 1–14 (2013)
8. G. Hariharan, An efficient wavelet based approximation method to water quality assessment model in a uniform channel. Ains Shams Eng. J. **5**, 525–532 (2014)
9. G. Hariharan, An efficient Legendre wavelet based approximation method for a few-Newell and Allen-Cahn equations. J. Membr. Biol. **247**(5), 371–380 (2014)

Chapter 4
Wavelet Method to Film–Pore Diffusion Model for Methylene Blue Adsorption onto Plant Leaf Powders

In this chapter, we have developed an accurate and efficient Haar wavelet method (HWM) to solve film–pore diffusion model. Film–pore diffusion model is widely used to determine study the kinetics of adsorption systems. To the best of our knowledge, until now rigorous wavelet solution has been not reported for solving film–pore diffusion model. The use of Haar wavelets is found to be accurate, simple, fast, flexible, convenient, and computationally attractive. The power of the manageable method is confirmed. It is shown that film–pore diffusion model satisfactorily describes kinetics of methylene blue adsorption onto the three low-cost adsorbents, Guava, teak, and gulmohar plant leaf powders, used in this study.

4.1 Introduction

Adsorption process has been proven to be one of the highly efficient methods for the removal of colors, odors, and organic and inorganic pollutants emanating from various industrial processes. Large amounts of dyes are used by textile industry and a significant portion of these dyes are not consumed in the process and therefore let out with the effluent. As the cost of commercial adsorbents is too high, interest for using low-cost adsorbents for removal of dyes from textile effluents is continuously growing. A recent survey indicates that, in India, on average freshwater consumed and effluent generated per kg of finished textile are 175 and 125 L, respectively [1]. The presence of dyes in aqueous effluents is highly objectionable as this affects the photosynthetic activity in receiving water body by reducing/preventing light penetration. As the dyes are recalcitrant in nature, it is difficult to treat them in conventional biological treatment plant [2, 3]. Various researchers have worked on biological degradation of dyes. But, very often, the metabolic intermediates are found to be more toxic than the original compound [4]. Therefore, identification of low-cost adsorbents is given more attention by the researchers recently as commercial adsorbents like activated carbon are too costly. Few recent studies

© Springer Nature Singapore Pte Ltd. 2019   51
G. Hariharan, *Wavelet Solutions for Reaction–Diffusion Problems in Science and Engineering*, Forum for Interdisciplinary Mathematics,
https://doi.org/10.1007/978-981-32-9960-3_4

investigating application of low-cost adsorbents are jackfruit peel [5], pineapple stem [6], phoenix tree leaves [7], pomelo peel [8], shells of bittim [9], orange peel [10], broad been peels [11], etc.

Adsorption of dye is complex process involving one or more of the following consecutive steps (i) diffusion of dye molecules across the external liquid film surrounding the solid particles, (ii) adsorption and desorption on the external surface of the particle, (iii) internal diffusion of dye within the particle either by pore diffusion, or surface diffusion or both, and (iv) adsorption and desorption on the internal surface of the particle. Since adsorption is a surface phenomena and majority of the adsorbents used are porous, external and internal resistances to the mass transfer of the solute play major role in controlling the rate of adsorption. In order to determine the rate controlling step and to understand the adsorption mechanism, it is necessary to determine external mass transfer coefficient and internal pore diffusivity. Simplified single resistance models are available to predict external film transfer coefficients. These are robust models, efficient for quick estimation of mass transfer parameters mentioned above. However, accurate values of the parameters can only be obtained using more rigorous two resistance models. Film–pore diffusion model (FPDM) was employed successfully to describe the kinetics of methylene blue adsorption onto GLP, TLP, and GUL. Diffusion-based kinetic models are too complex and require rigorous solution methods. For many of the diffusion models, pure analytical solution is not possible. In our previous paper, we had employed method of lines to solve film–pore diffusion model and had shown that film–pore model could describe the kinetics of adsorption of MB onto GLP, TLP, and GUL [1]. In this work, we have proposed a Haar solution to film–pore diffusion model. In our previous reports, we have established the feasibility and adsorption of MB onto three plant leaf powders namely guava leaf powder (GLP), teak leaf powder (TLP), and gulmohar leaf powder (GUL) [1].

As a powerful mathematical tool, wavelet analysis has been widely used in image digital processing, quantum field theory, numerical analysis, and many other fields in recent years. The Haar transform is one of the earliest examples of what is known now as a compact, dyadic, orthonormal wavelet transform. The Haar function, being an odd rectangular pulse pair, is the simplest and oldest orthonormal wavelet with compact support. In the meantime, several definitions of the Haar functions and various generalizations have been published and used. They were intended to adopt this concept to some practical applications as well as to extend its in applications to different classes of signals. Haar functions appear very attractive in many applications, as for example, image coding, edge extraction, and binary logic design.

Chen and Hsiao [12] first derived a Haar operational matrix for the integrals of the Haar function vector and demonstrated the application of Haar analysis in dynamic systems. Then Hsiao [13], who first proposed a Haar product matrix and a coefficient matrix, laid down the pioneer work in state analysis of linear time delayed systems via Haar wavelets. In order to take the advantages of the local property, several authors had used the Haar wavelet to solve the differential and integral equations [14–18]. Lepik [19–21] had solved higher order as well as nonlinear ODEs and some nonlinear evolution equations by Haar wavelet method. Hariharan et al. [22–25] have introduced the solution of Fisher's equation, Cahn–Allen equation, convection–diffusion equations, and some nonlinear parabolic equations by the Haar wavelet method. The fundamental idea of Haar wavelet method is to convert the problem of solving for the one-dimensional differential equation with constant coefficients, which satisfies the boundary conditions and initial conditions into a group of algebraic equations, which involves a finite number of variables.

In this chapter, we establish a clear procedure for solving the differential equations with constant coefficients via Haar wavelet. The Haar wavelet is introduced, and an operational matrix is established first, and then a direct method for solving the differential equations via Haar wavelet is demonstrated. Because of the local property of Haar wavelet, the new method is simpler in reasoning as well as in calculation.

4.2   Materials and Methods

Detailed development of FPDM is described earlier by McKay and co-workers [19, 20]. Solution of FPDM by method of lines is described in our previous paper [1]. In the present chapter, development of Haar solution is described in detailed and the results are compared with our previous solution.

4.3   Haar Wavelet and Its Properties

Haar wavelet was a system of square wave; the first curve was marked up as $h_0(t)$, the second curve marked up as $h_1(t)$ that is

$$h_0(x) = \begin{cases} 1, & 0 \leq x < 1 \\ 0, & \text{otherwise} \end{cases} \tag{4.1}$$

$$h_1(x) = \begin{cases} 1, & 0 \leq x < 1/2, \\ -1, & 1/2 \leq x < 1, \\ 0, & \text{otherwise}, \end{cases} \tag{4.2}$$

where $h_0(x)$ is scaling function, $h_1(x)$ is mother wavelet. In order to perform wavelet transform, Haar wavelet uses dilations and translations of function, i.e., the transform makes the following function.

$$h_n(x) = h_1\left(2^j x - k\right), \ n = 2^j + k, \ j \geq 0, \ 0 \leq k < 2^j. \tag{4.3}$$

4.3.1 Function Approximation

Any square-integrable function $y(x) \in L^2[0, 1)$ can be expanded by a Haar series of infinite terms

$$y(x) = \sum_{i=0}^{\infty} c_i h_i(x), \ i \in \{0\} \cup N, \tag{4.4}$$

where the Haar coefficients c_i are determined as, $c_0 = \int_0^1 y(x) h_0(x) dx,$

$c_n = 2^j \int_0^1 y(x) h_i(x) dx, \ i = 2^j + k, \ j \geq 0, \ 0 \leq k < 2^j, \ x \in [0, 1)$ such that the following integral square error ε is minimized:

$$\varepsilon = \int_0^1 \left[y(x) - \sum_{i=0}^{m-1} c_i h_i(x) \right]^2 dx, \ m = 2^j, \ j \in \{0\} \cup N. \tag{4.5}$$

Usually, the series expansion contains infinite terms for smooth $y(x)$. If $y(x)$ is piecewise constant by itself or may be approximated as piecewise constant during each subinterval, then $y(x)$ will be terminated at finite m terms, that is

$$y(x) = \sum_{i=0}^{m-1} c_i h_i(x) = c_{(m)}^T h_{(m)}(x), \tag{4.6}$$

where the coefficients $c_{(m)}^T$ and the Haar function vector $h_{(m)}(x)$ are defined as $c_{(m)}^T = [c_0, c_1, \ldots, c_{m-1}]$ and $h_{(m)}(x) = [h_0(x), h_1(x), \ldots, h_{m-1}(x)]^T$ where 'T' means transpose and $m = 2^j$.

The first four Haar function vectors, which $x = n/8, \ n = 1, 3, 5, 7$ can be expressed as follows:

$$h_4(1/8) = [1, 1, 1, 0]^T, \quad h_4(3/8) = [1, 1, -1, 0]^T,$$
$$h_4(5/8) = [1, -1, 0, 1]^T, \ h_4(7/8) = [1, -1, 0, -1]^T,$$

which can be written in matrix form as

$$H_4 = [h_4(1/8), h_4(3/8), h_4(5/8), h_4(7/8)]$$

$$H_4 = \begin{bmatrix} 1 & 1 & 1 & 1 \\ 1 & 1 & -1 & -1 \\ 1 & -1 & 0 & 0 \\ 0 & 0 & 1 & -1 \end{bmatrix},$$

In general, we have

$$H_m = [h_m(1/2m), h_m(3/2m), \dots, h_m(2m-1)/2m],$$

where $H_1 = [1]$, $H_2 = \begin{pmatrix} 1 & 1 \\ 1 & -1 \end{pmatrix}$. The collocation points are identified as $x_l = (2l-1)/2m$, $l = 1, 2, \dots, m$. In application, in order to avoid dealing with impulse function, integration of the vector $h_m(x)$ given by

$$\int_0^x h_m(t)dt \approx P_m h_m(x), \ x \in [0, 1], \tag{4.7}$$

where P_m is the $m \times m$ operational matrix and is given by

$$P_{(m)} = \frac{1}{2m} \begin{pmatrix} 2mP_{(m/2)} & -H_{(m/2)} \\ H_{(m/2)}^{-1} & O \end{pmatrix}$$

where O is a null matrix of order $\frac{m}{2} \times \frac{m}{2}$.
  Here $P_1 = [1/2]$, so

$$P_2 = \frac{1}{4} \begin{pmatrix} 2 & -1 \\ 1 & 0 \end{pmatrix}, \quad P_4 = \frac{1}{16} \begin{bmatrix} 8 & -4 & -2 & -2 \\ 4 & 0 & -2 & 2 \\ 1 & 1 & 0 & 0 \\ 1 & -1 & 0 & 0 \end{bmatrix},$$

$$P_8 = \frac{1}{64} \begin{bmatrix} 32 & -16 & -8 & -8 & -4 & -4 & -4 & -4 \\ 16 & 0 & -8 & 8 & -4 & -4 & 4 & 4 \\ 4 & 4 & 0 & 0 & -4 & 4 & 0 & 0 \\ 4 & 4 & 0 & 0 & -4 & 4 & 0 & 0 \\ 1 & 1 & 2 & 0 & 0 & 0 & 0 & 0 \\ 1 & 1 & -2 & 0 & 0 & 0 & 0 & 0 \\ 1 & -1 & 0 & 2 & 0 & 0 & 0 & 0 \\ 1 & -1 & 0 & -2 & 0 & 0 & 0 & 0 \end{bmatrix}$$

It should be noted that calculations for $P_{(m)}$ and $H_{(m)}$ must be carried out only once; after that they will be applicable for solving whatever differential equations. The fast capability of HT should be impressive. Since H and H^{-1} contain many zeros, this phenomenon makes the Haar transform faster than the Fourier transform, and it is even faster than the Walsh transform. This is one of the reasons for rapid convergence of the Haar wavelet series. The numbers of additions and multiplications for these three transforms are shown in Table 4.1.

In practical applications, a small number of terms increases the calculation speed and saves memory storage; a large number of terms improve resolution accuracy. Therefore, a trade-off between calculation speed, memory saving, and the resolution accuracy has been considered in the analysis.

4.4 Method of Solution

Consider the equation

$$\dot{\bar{C}}_i(Z, \tau) = A(\bar{C}_i) \left[\bar{C}_i'' + \left(\frac{1}{Z} \right) \bar{C}_i' \right] \tag{4.8}$$

Let us divide the interval $(0,1]$ into N equal parts of length $\Delta\tau = (0, 1]/N$ and denote $\tau_s = (s - 1)\Delta t$, $s = 1, 2, \ldots N$. We assume that $\dot{\bar{C}}''(Z, \tau)$ can be expanded in terms of Haar wavelets as formula

$$\dot{\bar{C}}_i''(Z, \tau) = \sum_{i=0}^{m-1} c_s(i) h_i(Z) = c_{(m)}^{\mathrm{T}} h_{(m)}(Z) \tag{4.9}$$

where . and $'$ means differentiation with respect to t and x, respectively, the row vector $c_{(m)}^{\mathrm{T}}$ is constant in the subinterval $\tau \in (\tau_s, \tau_{s+1}]$

Integrating Formula (4.9) with respect to τ from τ_s to τ and twice with respect to Z from 0 to x, we obtain

$$\bar{C}_i''(Z, \tau) = (\tau - \tau_s) c_{(m)}^{\mathrm{T}} h_{(m)}(Z) + \bar{C}_i''(Z, \tau_s) \tag{4.10}$$

$$\begin{aligned} \bar{C}_i(Z, \tau) = (\tau - \tau_s) c_{(m)}^{\mathrm{T}} Q_{(m)} h_{(m)}(Z) + \bar{C}_i(Z, \tau_s) - \bar{C}_i(0, \tau_s) \\ + Z[\bar{C}_i'(0, \tau) - \bar{C}_i'(0, \tau_s)] + \bar{C}_i(0, \tau) \end{aligned} \tag{4.11}$$

$$\bar{C}_i'(Z, \tau) = (\tau - \tau_s) c_{(m)}^{\mathrm{T}} P_{(m)} h_{(m)}(Z) + \bar{C}_i'(Z, \tau_s) - \bar{C}_i(0, \tau_s) + \bar{C}_i'(0, \tau) \tag{4.12}$$

$$\dot{\bar{C}}_i(Z, \tau) = c_{(m)}^{\mathrm{T}} Q_{(m)} h_{(m)}(Z) + Z\dot{\bar{C}}_i(0, \tau) + \dot{\bar{C}}_i(0, \tau) \tag{4.13}$$

with the boundary conditions, we obtain

$$\begin{aligned} \bar{C}_i(0, \tau_s) = g_0(\tau_s), \quad \bar{C}_i(1, \tau_s) = g_1(\tau_s) \\ \dot{\bar{C}}_i(0, \tau) = g_0'(\tau), \quad \dot{\bar{C}}_i(1, \tau) = g_1'(\tau) \end{aligned}$$

Table 4.1 Comparison between Haar wavelet method (HWM) and method of lines (MOL) by obtaining the mass transfer coefficients using film–pore diffusion model adsorption of MB onto GLP and $m = 16$, $t = 10$ s

Temperature (K)	$C_0\,(mg\,dm^{-3})$	$k_f\,(ms^{-1})$		$D_{\text{eff}}\,(m^2s^{-1})$		Error	
		MOL (M)	HWM (H)	MOL (M)	HWM (H)	E_M	E_H
303	50	1.00×10^{-6}	3.00×10^{-6}	1.74×10^{-13}	2.54×10^{-13}	1.197	0.938
	100					0.140	0.129
	150					0.935	0.824
	200					1.610	1.102
313	50	1.71×10^{-6}	3.45×10^{-6}	6.46×10^{-13}	9.26×10^{-13}	1.462	1.221
	100					1.120	0.927
	150					1.267	1.016
	200					7.570	5.112
323	50	4.27×10^{-6}	4.75×10^{-6}	3.11×10^{-13}	5.31×10^{-13}	0.856	0.284
	100					0.160	0.000
	150					2.168	1.208
	200					3.164	1.016

E_M Error by method of lines
E_H Error by Haar wavelet method

Putting $x = 1$ in Formulae (4.10)–(4.13), we have

$$\bar{C}'_i(0, \tau) - \bar{C}'_i(0, \tau_s) = -(\tau - \tau_s)c^{\mathrm{T}}_{(m)}P_{(m)}h_{(m)}(Z) + g_1(\tau) - g_0(\tau) - g_1(\tau_s) + g_0(\tau_s)$$

$$(4.14)$$

$$\dot{\bar{C}}'_i(0, \tau) = g'_1(\tau) - c^{\mathrm{T}}_{(m)}Q_{(m)}h_{(m)}(Z)f - g'_0(\tau)$$
$$(4.15)$$

where the vector f is defined as

$$f = [1, \underbrace{0, \ldots, 0}_{(m-1)\ \text{elements}}]^T$$

Substituting Formulae (4.14) and (4.15) into Formulae (4.10)–(4.13), and discretizising the results by assuming $Z \to Z_l$, $\tau \to \tau_{s+1}$, we obtain

$$\bar{C}''_i(Z_l, \tau_{s+1}) = (\tau_{s+1} - \tau_s)c^{\mathrm{T}}_{(m)}h_{(m)}(Z_l) + \bar{C}''_i(Z_l, \tau_s)$$
$$(4.16)$$

$$\bar{C}'_i(Z_l, \tau_{s+1}) = (\tau_{s+1} - \tau_s)c^{\mathrm{T}}_{(m)}Q_{(m)}h_{(m)}(Z_l) + \bar{C}_i(Z_l, \tau_s) - g_0(\tau_s) + g_0(\tau_{s+1})$$
$$+ Z_l[-(\tau_{s+1} - \tau_s)c^{\mathrm{T}}_{(m)}P_{(m)}f + g_l(\tau_{s+1}) - g_0(\tau_{s+1}) - g_1(\tau_s) + g_0(\tau_s)]$$
$$(4.17)$$

$$\dot{\bar{C}}_i(Z_l, \tau_{s+1}) = c^{\mathrm{T}}_{(m)}Q_{(m)}h_{(m)}(Z) + Z\dot{\bar{C}}_i(0, \tau) + \dot{\bar{C}}_i(0, \tau)$$
$$(4.18)$$

$$\dot{\bar{C}}_i(Z_l, \tau_{s+1}) = c^{\mathrm{T}}_{(m)}Q_{(m)}h_{(m)}(Z) + g'_0(\tau_{s+1}) + Z_l[-c^{\mathrm{T}}_{(m)}P_{(m)}f + g'_1(\tau_{s+1})$$
$$- g'_0(\tau_{s+1})]$$
$$(4.19)$$

There are several possibilities for treating the nonlinearity in Eq. (4.10). In the following, the scheme

$$\dot{\bar{C}}_i(Z_l, \tau_{s+1}) = A(\bar{C}_i)\left[\bar{C}''_i(Z_l, \tau_{s+1}) + \frac{1}{Z}\bar{C}'_i(Z_l, \tau_{s+1})\right]$$
$$(4.20)$$

which leads us from the time layer τ_s to τ_{s+1} is used.

Substituting Formulae (4.16)–(4.19) into Formula (4.20), we gain

$$c^{\mathrm{T}}_{(m)}Q_{(m)}h_{(m)}(Z_l) + Z_l\left[-c^{\mathrm{T}}_{(m)}P_{(m)}f + g'_1(\tau_{s+1}) - g'_0(\tau_{s+1})\right] + g'_0(\tau_{s+1})$$
$$= A(\bar{C}_i)\begin{bmatrix}(\tau_{s+1} - \tau_s)c^{\mathrm{T}}_{(m)}h_{(m)}(Z_l) + (\tau_{s+1} - \tau_s)c^{\mathrm{T}}_{(m)}Q_{(m)}h_{(m)}(Z_l) + \bar{C}_i(Z_l, \tau_s) - g_0(\tau_s) + g_0(\tau_{s+1}) \\ + Z_l[-(\tau_{s+1} - \tau_s)c^{\mathrm{T}}_{(m)}P_{(m)}f + g_l(\tau_{s+1}) - g_0(\tau_{s+1}) - g_1(\tau_s) + g_0(\tau_s)]\end{bmatrix}$$

From the above formula, the wavelet coefficients $c^{\mathrm{T}}_{(m)}$ can be successively calculated.

Here $A(\bar{C}_i)$ are constants (linear) and $\in = 0.5$, $\rho = 500$.

Table 4.1 gives a comparison of Haar wavelet solutions and method of lines. It is evident that Haar wavelet solutions are better than that of the method of lines. Value of absolute error decreased when m was increased. The results show that combining with wavelet matrix, the method in this paper can be effectively used in numerical calculus for constant coefficient differential equations, and that the method is feasible. At the same time with the sparse nature of Haar wavelet matrix, compared with the method of lines [1], using the above method can greatly reduce the computation and from the above results, we can see that the numerical solutions are in good agreement with exact solution. The power of the manageable method is thus confirmed.

All the numerical experiments presented in this section were computed in double precision with some MATLAB codes on a personal computer System with Processor Intel(R) Core$^{(TM)}$ 2 Duo CPU T5470 @ 1.60 GHz(2CPUs) and 1 GB RAM.

4.5 Conclusion

In the present chapter, FPDM model equations had been solved by Haar wavelet method. It was found that the model could predict the concentration decay curve for all adsorption of methylene blue onto TLP, GUL, and GLP excellently with a small deviation during initial period. The theoretical elegance of the Haar wavelet approach can be appreciated from the simple mathematical relations and their compact derivations and proofs. It has been well demonstrated that in applying the nice properties of Haar wavelets, the differential equations can be solved conveniently and accurately by using Haar wavelet method systematically. According to this method, the spatial operators are approximated by the Haar wavelet method and the time derivation operators by the finite difference method. The main advantage of this method is its simplicity and small computation costs. It is due to the sparsity of the transform matrices and to the small number of significant wavelet coefficients. In comparison with existing numerical schemes used to solve the nonlinear parabolic equations, the scheme in this paper is an improvement over other methods in terms of accuracy. It is worth mentioning that Haar solution provides excellent results even for small values of m ($m = 16$). For larger values of m (ie., $m = 32$, $m = 64$, $m = 128$, $m = 256$), we can obtain the results closer to the real values. The method with far less degrees of freedom and with smaller CPU time provides better solutions than classical ones. The work confirmed the power of the Haar wavelet method in handling nonlinear equations. This method can be easily extended to find the solution of all other nonlinear parabolic equations too.

References

1. V. Ponnusami, K.S. Rajan, S.N. Srivastava, Application of film-pore diffusion model, for methylene blue adsorption onto plant leaf powders. Chem. Eng. J. **163**(3), 236–242 (2010)
2. V. Ponnusami, V. Gunasekar, S.N. Srivastava, Kinetics of methylene blue removal from aqueous solution using gulmohar (Delonix regia) plant leaf powder: Multivariate regression analysis. J. Hazard. Mater. **169**, 119–127 (2009)
3. V. Ponnusami, V. Krithika, R. Madhuram, S.N. Srivastava, Biosorption of reactive dye using acid treated rice husk: Factorial design analysis. J. Hazard. Mater. **142**, 397–403 (2007)
4. C.W. Cheung, C.K. Chan, J.F. Porter, G. McKay, Film-pore diffusion control for the batch sorption of cadmium ions from effluent onto bone char. J. Colloid Interface Sci. **234**(2), 328–336 (2001)
5. B.H. Hameed, Removal of cationic dye from aqueous solution using jackfruit peel as non-conventional low-cost adsorbent. J. Hazard. Mater. **162**(1), 344–350 (2009)
6. B.H. Hameed, R.R. Krishni, S.A. Sata, A novel agricultural waste adsorbent for the removal of cationic dye from aqueous solutions. J. Hazard. Mater. **162**(1), 305–311 (2009)
7. N. Gupta, A.K. Kushwaha, M.C. Chattopadhyaya, Adsorption studies of cationic dyes onto Ashoka (Saraca asoca) leaf powder. J. Taiwan Inst. Chem. Eng. **43**(4), 604–613 (2012)
8. B.H. Hameed, D.K. Mahmoud, A.L. Ahmad, Sorption of basic dye from aqueous solution by pomelo (Citrus grandis) peel in a batch system. Colloids Surf. A Physicochem. Eng. Aspects **316**(1–3), 78–84 (2008)
9. A. Çelekli, G. Ilgün, H. Bozkurt, Sorption equilibrium, kinetic, thermodynamic, and desorption studies of reactive red 120 on Chara contraria. Chem. Eng. J. **191**, 228–235 (2012)
10. S. Kumar, V. Gunasekar, V. Ponnusami, Removal of methylene blue from aqueous effluent using fixed bed of groundnut shell powder. J. Chem. (in press). https://doi.org/10.1155/2013/259819
11. B.H. Hameed, M.I. El-Khaiary, Sorption kinetics and isotherm studies of a cationic dye using agricultural waste: broad bean peels. J. Hazard. Mater. **154**(1–3), 639–648 (2008)
12. C.F. Chen, C.H. Hsiao, Haar wavelet method for solving lumped and distributed-parameter systems. IEEE Proc. Pt. D **144**(1), 87–94 (1997) 123 J Math Chem (2012) **50**:2775–2785 2785
13. C.H. Hsiao, State analysis of linear time delayed systems via Haar wavelets. Math. Comput. Simul. **44**(5), 457–470 (1997)
14. Z. Shi, T. Liu, B. Gao, Haar wavelet method for solving wave equation. in *International Conference on Computer Application and System Modeling (ICCASM 2010), IEEE Proceeding*. (2010)
15. F.I. Haq, I. Aziz, S.U. Islam, A Haar wavelets based numerical method for eight-order boundary problems. Int. J. Math. Comput. Sci. **6**(1), 25–31 (2010)
16. J.L. Wu, A wavelet operational method for solving fractional partial differential equations numerically. Appl. Math. Comput. **214**(1), 31–40 (2009)
17. Z. Shi, Y.-Y. Cao, Q.-J. Chen, Solving 2D and 3D Poisson equations and biharmonic equations by the Haar wavelet method. Appl. Math. Model. **36**(11), 5143–5161 (2012)
18. W. Geng, Y. Chen, Y. Li, D. Wang, Wavelet method for nonlinear partial differential equations of fractional order. Comput. Inf. Sci. **4**(5), 28–35 (2011)
19. U. Lepik, Numerical solution of evolution equations by the Haar wavelet method. Appl. Math. Comput. **185**, 695–704 (2007)
20. U. Lepik, Numerical solution of differential equations using Haar wavelets. Math. Comput. Simul. **68**, 127–143 (2005)
21. U. Lepik, Application of the Haar wavelet transform to solving integral and differential equations. Proc. Estonian Acad. Sci. Phys. Math. **56**(1), 28–46 (2007)
22. G. Hariharan, K. Kannan, Haar wavelet method for solving some nonlinear parabolic equations. J. Math. Chem. **48**(4), 1044–1061 (2010)

23. G. Hariharan, K. Kannan, A comparative study of a Haar Wavelet method and a restrictive Taylor's series method for solving convection-diffusion equations. Int. J. Comput. Methods Eng. Sci. Mech. **11**(4), 173–184 (2010)
24. G. Hariharan, Haar wavelet method for solving Sine-Gordon and Klein-Gordon equations. Int. J. Nonlinear Sci. **9**(2), 1–10 (2010)
25. G. Hariharan, K. Kannan, Haar wavelet method for solving fitz Hugh-Nagumo equation. Int. J. Math. Stat. Sci. **2**, 2 (2010)

Chapter 5
An Efficient Wavelet-Based Spectral Method to Singular Boundary Value Problems

In this chapter, an efficient wavelet-based approximation method is established to nonlinear singular boundary value problems. To the best of our knowledge, until now there is no rigorous shifted second kind Chebyshev wavelet (S2KCWM) solution has been addressed for the nonlinear differential equations in population biology. With the help of shifted second kind Chebyshev wavelet operational matrices, the nonlinear differential equations are converted into a system of algebraic equations. The convergence of the proposed method is established. The power of the manageable method is confirmed. Finally, we have given some numerical examples to demonstrate the validity and applicability of the proposed wavelet method.

5.1 Introduction

In recent years, nonlinear singular boundary value problems (NSBVPs) arise in many branches of engineering and applied mathematics such as chemical reactions gas dynamics, electro-hydrodynamics, nuclear physics, atomic structures, atomic calculations, physiology, and medical sciences. The numerical study of singular boundary value problems arising in various physical models has been done by many authors [1–11], and a variety of methods have been introduced to solve such singular boundary value problems. [2, 3, 5, 8–10]. Although, these numerical methods have many advantages, a huge amount of computational work is needed.

Wavelet-based spectral methods have been successfully introduced for solving nonlinear-type differential equations from the beginning of 1990s. In the last few decades, the wavelet-based approximation methods for such problems have attracted excellent attention and numerous papers about this topic have been published. Wavelet analysis possesses many useful properties, such as compact support, orthogonality, dyadic, orthonormality, and multi-resolution analysis (MRA). An excellent discussion on wavelet transforms and the Fourier transforms presented

© Springer Nature Singapore Pte Ltd. 2019
G. Hariharan, *Wavelet Solutions for Reaction–Diffusion Problems in Science and Engineering*, Forum for Interdisciplinary Mathematics,
https://doi.org/10.1007/978-981-32-9960-3_5

by Gilbert Strang in the year 1993. In the numerical analysis, wavelet-based methods and hybrid methods become important tools because of the properties of localization. In wavelet-based techniques, there are two important ways of improving the approximation of the solutions: increasing the order of the wavelet family and increasing the resolution level of the wavelet. There is a growing interest in using wavelets to study problems of greater computational complexity. Wavelet methods have proved to be very effective and efficient tool for solving problems of mathematical calculus [10–29]. Among the wavelet transform families the Haar, Legendre wavelets and Chebyshev wavelets deserve much attention [25–29].

The basic idea of Chebyshev wavelet method (CWM) is to convert the differential equations into a system of algebraic equations by the operational matrices of integral or derivative. The main goal is to show how wavelets and multi-resolution analysis (MRA) can be applied for improving the method in terms of easy implementability and achieving the rapidity of its convergence. Wavelets, as very well-localized functions, are considerably useful for solving differential equations and provide accurate solutions. Also, the wavelet technique allows the creation of very fast algorithms when compared with the algorithms ordinarily used [25–28]. Recently, Hariharan and Kannan [29] reviewed the wavelet transform methods for solving a few reaction–diffusion equations arising in science and engineering. In this paper, the shifted second kind Chebyshev wavelet method (S2KCWM) is applied to singular boundary value problems arising in biology. The method consists of reducing the differential equations to a set of algebraic equations by first expanding the candidate function as Chebyshev wavelets with unknown coefficients [12–25].

Consider Lane–Emden-type equation

$$\frac{d^2 y}{dx^2} + \frac{\alpha}{x}\frac{dy}{dx} = f(x, y), 0 \le x \le 1, \alpha \ge 1, \tag{5.1}$$

with the boundary conditions

$$y'(0) = 0, \quad ay(1) + by'(1) = c, \tag{5.2}$$

where a, b, and c are real constants.

The above equation represents the heat conduction through a solid with heat generation source within the solid. The function $f(x, y)$ represents the heat generation within the solid. If $\alpha = 0$, the solid is a plane. If $\alpha = 1$, the solid is a cylinder. If $\alpha = 2$, the solid is a sphere.

5.2   Nonlinear Stability Analysis of Lane–Emden Equation of First Kind (Emden–Fowler Equation)

This equation arises in the study of stellar interiors and also is called the polytropic differential equation.

For polytropic perfect fluid sphere, the Emden–Fowler equation in $\theta(\xi)$ is given by

$$\frac{d^2\theta}{d\xi} + \frac{2}{\xi}\frac{d\theta}{d\xi} + \theta^n = 0, \tag{5.3}$$

where ξ is a dimensionless radial coordinate and θ is related to the density and thus the pressure by $\rho = \rho_c \theta^n$ where ρ_c is the central density. The index n is known as a polytropic index. It appears in the polytropic equation of the state $P = K\rho^{1+\frac{1}{n}}$, P is the pressure, and ρ is density, and K is the constant of proportionality.

The boundary conditions are

$$\theta'(0) = 0, \ \theta(0) = 1. \tag{5.4}$$

5.2.1   The Emden–Fowler Equation as an Autonomous System

In order to study the stability of the equilibrium points of the Emden–Fowler, we rewrite it in the form of an autonomous system of differential equations.

Theorem 5.1 Emden–Fowler Eq. (5.3) is equivalent to the following autonomous system of differential equations

$$\frac{dw}{dt} = q, \tag{5.5}$$

$$\frac{dq}{dt} = -2G^1(w, q), \tag{5.6}$$

with the function $G^1(w, q)$ is given by

$$G^1(w, q) = \frac{1}{2}\left[-\frac{n-5}{n-1}q + \frac{2(3-n)}{(n-1)^2}w + B^{n-1}w^n\right]. \tag{5.7}$$

The set of new variables (w, t) is defined as $\theta(\xi) = B\xi^{\frac{2}{(1-n)}}w(\xi)$, $\xi = \xi_s e^{-t}$ where ξ_s is the value of ξ at star's surface, and $B > 0$ is a constant. The range of radial dimensionless variable ξ is

$\xi \in [0, \xi_s]$. The range of new variable $t = \ln\left(\frac{\xi_s}{\xi}\right)$ is $t \in (\infty, 0]$.

Proof The critical points of this dynamical system all reside on the $q = 0$ line. They are given the solutions of the equation

$$G^1(w, 0) = \frac{1}{2} \left[-\frac{2(n-3)}{(n-1)^2} + B^{n-1} w^n \right] = 0. \tag{5.8}$$

Therefore the critical points $X_i = (w_0, q_0)$ of the systems (5.5) and (5.6), we firstly find the point

$$X_0 = (0, 0), \tag{5.9}$$

whose value is n independent.

For $w \neq 0$, Eq. (5.8) is equivalent to

$$\frac{2(n-3)}{(n-1)^2} = B^{n-1} w^{n-1}. \tag{5.10}$$

Hence for $1 < n < 3$, we find

$$X_n = \left(\sqrt[n]{-1} \frac{1}{B} \sqrt[n-1]{\frac{2(3-n)}{(n-1)^2}}, 0 \right), \tag{5.11}$$

and for $n > 3$, we have

$$X_n = \left(\frac{1}{B} \left[\frac{2(n-3)}{(n-1)^2} \right]^{\frac{1}{n-1}}, 0 \right). \tag{5.12}$$

Definition 5.1

(Lyapunov function). Given a smooth dynamical system $\dot{x} = f(x), x \in R^n$, and an equilibrium point x_0, a continuous function $V : R^n \rightarrow R$ in a neighborhood U of x_0 is a Lyapunov function for the point x_0 if

1. V is differentiable in $U \backslash x_0$.
2. $V(x) > V(x_0)$.
3. $\dot{V}(x) \leq 0$ for every $x \in U \backslash x_0$.

The neighborhood U is constrained by the third condition. When '3' holds, the method provides information not only about the asymptotic stability of the equilibrium point but also about its basin of attraction, which must contain the set U. The existence of a Lyapunov function V for an autonomous system of differential equations guarantees the stability of the point x_0. This qualitative result can be obtained without explicitly solving the equations, and this information cannot be achieved by using linear stability analysis or first-order perturbation theory. The following theorem holds:

Theorem 5.2 (Lyapunov stability). Let x_0 be an equilibrium point of the system $\dot{x} = f(x)$, where $f : U \in \mathfrak{R}^n$ is locally Lipschitz, and $U \subset R^n$ is a domain that contains x_0. If V is a Lyapunov function, then

(1) $\dot{V}(x) = \frac{\partial V}{\partial x}, f$ is negative semi-definite, then $x = x_0$ is a stable equilibrium point,
(2) $\dot{V}(x) = \frac{\partial V}{\partial x}, f$ is negative definite, then $x = x_0$ is an asymptotically stable equilibrium point.

Moreover, if $\|x\| \to \infty$ implies that $V(x) \to \infty$ for all x, then x_0 is globally stable or globally asymptotically stable, respectively. If the condition 3 of the Lyapunov function definition holds strictly, the existence of a Lyapunov function has important consequences for the behavior of the time-dependent perturbations of autonomous systems.

5.2.2 Application of Stability Analysis to Emden–Fowler Equation

Theorem 5.3 Let the dynamical system (5.5) and (5.6), with critical points (5.9) and (5.12), be given. The equilibrium state X_0 is asymptotically stable is $1 < n < 3$, and the equilibrium state X_n is asymptotically stable is $3 < n < 5$.

Proof One possible Lyapunov function $V(w, q)$ associated with the system given by Eqs. (5.5) and (5.6) is chosen [16]. Following the variable gradient method (this means setting the $\nabla V = f \in \mathfrak{R}^n$), We set

$$\nabla V(w, q) = \begin{pmatrix} -2(n-3)/(n-1)^2 w + B^{n-1} w^n \\ q \end{pmatrix}, \tag{5.13}$$

such that the critical points correspond to $\nabla V = 0$(since this corresponds to $f = 0$). It yields the Lyapunov function

$$V(w, q) = \frac{1}{2}q^2 - \frac{(n-3)}{(n-1)^2}w^2 + \frac{B^{n-1}}{n+1}w^{n+1}. \tag{5.14}$$

By definition, the Lyapunov function must have a local minimum at the critical points. To check this, we consider the Hessian of (5.14) which is given by

$$H(V) = \begin{pmatrix} -2(n-3)/(n-1)^2 + B^{n-1}nw^{n-1} & 0 \\ 0 & 1 \end{pmatrix}, \tag{5.15}$$

and has the eigenvalues

$$\lambda_1 = -2(n-3)/(n-1)^2 + B^{n-1}nw^{n-1}, \lambda_2 = 1. \tag{5.16}$$

At the point $X_0 = (0,0)$ for $n > 1$, we have

$$\lambda_1 = -2(n-3)/(n-1)^2, \lambda_2 = 1. \tag{5.17}$$

And hence there is a local minimum near X_0 if $1 < n < 3$.

At the point X_n, we find

$$\lambda_1 = 2(n-3)/(n-1), \lambda_2 = 1. \tag{5.18}$$

And hence there is a local minimum near X_n if $n > 3$ and this function has no local minimum if the index satisfies $1 < n < 3$.

The Lyapunov function (5.14) satisfies

$$\frac{dV}{dt} = \frac{\partial V}{\partial W}\frac{dw}{dt} + \frac{\partial V}{\partial q}\frac{dq}{dt} = \frac{n-5}{n-1}q^2, \tag{5.19}$$

and therefore, we find that

$$\dot{V} < 0, \quad 1 < n < 5. \tag{5.20}$$

Hence according to Lyapunov theorem, the equilibrium state X_0 is asymptotically stable equilibrium points for $1 < n < 3$ while the equilibrium state X_n is asymptotically stable equilibrium points for $3 < n < 5$. Global asymptotic stability results cannot be inferred from our Lyapunov function.

5.3 Order of the S2KCW Method [25]

Shifted second kind Chebyshev wavelets are constructed using shifted second kind Chebyshev polynomials. They are denoted by $\psi_{nm}(t) = \psi(k, n, m, t)$ and have four arguments: K, n—any positive integer, m—order of second kind Chebyshev polynomials, t—normalized time. They are defined in the interval [0, 1] by

$$\psi_{nm}(t) = \begin{cases} \frac{\beta_m 2^{\frac{k}{2}}}{\sqrt{\pi}} U_m^*(2^{k+1}t - 2n - 1), & t \in \left[\frac{n}{2^k}, \frac{n+1}{2^k}\right] \\ 0, & \text{Otherwise,} \end{cases} \tag{5.21}$$

$m = 0, 1, \ldots M, \ n = 0, 1, \ldots 2^k,$

where

$$\beta_m = \begin{cases} \sqrt{2}, & m = 0, \\ 2, & m > 0. \end{cases} \tag{5.22}$$

Theorem 5.4 A function $f(t)$ is defined in $[0,1]$ with bounded second derivative $|f'(t)| \leq M_1$ and $|f''(t)| \leq M_2$ can be expanded as an infinite sum of S2KCWs, and the series converges uniformly to $f(t) = \sum_{n=1}^{\infty} \sum_{m=0}^{\infty} C_{nm} \psi_{nm}(t)$,

where

$C_{nm} = (f(t), \psi_{nm}(t))_{w_n} = \int_0^1 f(t) \psi_{nm}(t) w_n(t) dt$ in which $(.,.)$ denotes inner product in $L^2_{w_n}[0,1]$.

Proof We consider the following case:

For $m = 0$ and $n = 1,2, \ldots 2^k$, the S2KCWs form an orthonormal system on $[0,1]$ as:

$$\psi_{n0}(t) = \begin{cases} \sqrt{\frac{2}{\pi}} 2^{\frac{k}{2}}, & t \in \left[\frac{n}{2^k}, \frac{n+1}{2^k+1}\right], \\ 0, & \text{otherwise,} \end{cases} \tag{5.23}$$

By expanding the function $f(t)$ in terms of the basis function on $[0,1]$ we have

$$f(t) = \sum_{n=1}^{\infty} C_{n0} \psi_{n0}(t), \tag{5.24}$$

where

$$C_{n0} = \sqrt{\frac{2}{\pi}} 2^{\frac{k}{2}} \int_{\frac{n}{2^k}}^{\frac{n+1}{2^k}} f(t) w_n(t) dt. \tag{5.25}$$

by truncating the series in (5.24), it can be written as

$$\tilde{f}(t) \simeq \sum_{n=1}^{2^k} C_{n0} \psi_{n0}(t). \tag{5.26}$$

Theorem 5.5 Suppose $\tilde{f}(t)$ be the truncated expansion of the function $f(t)$ in the above basis function and $\tilde{e}(t) = \tilde{f}(t) - f(t)$ be the corresponding error. Then the expansion will converge in the sense that $\tilde{e}(t)$ approaches zero with the same rate as 2^k, i.e., $\|\tilde{e}(t)\| = O\left(\frac{1}{2^k}\right)$.

Proof By defining the error between $f(t)$ and its expansion over every subinterval as follows:

$$e_n(t) = C_{n0}\psi_{n0}(t) - f(t), \; x \in \left[\frac{n}{2^k}, \frac{n+1}{2^k}\right], \; n = 1, 2. \ldots 2^k, \qquad (5.27)$$

we have

$$
\begin{aligned}
\|e_n(t)\|^2 &= \int_{\frac{n}{2^k}}^{\frac{n+1}{2^k}} e_n(t)^2 w_n(t) dt = \int_{\frac{n}{2^k}}^{\frac{n+1}{2^k}} (C_{n0}\psi_{n0}(t) - f(t))^2 w_n(t) dt \\
&= \left(C_{n0}\sqrt{\frac{2}{\pi}}2^{\frac{k}{2}} - f(\eta_n)\right)^2 \frac{\pi}{2^{k+1}}, \; \eta_n \in \left[\frac{n}{2^k}, \frac{n+1}{2^k}\right],
\end{aligned}
\qquad (5.28)
$$

Where we have used the weighted mean value theorem for integrals. From (5.25), and the weighted mean value theorem, we also have

$$C_{n0} = \sqrt{\frac{2}{\pi}}2^k \int_{\frac{n}{2^k}}^{\frac{n+1}{2^k}} f(t)w_n(t) dt = \sqrt{\frac{2}{\pi}}2^k \frac{\pi}{2^{k+1}} f(\zeta_n), \; \zeta_n \in \left[\frac{n}{2^k}, \frac{n+1}{2^k}\right]. \qquad (5.29)$$

By substituting (5.29) into (5.28), we obtain

$$\|e_n(t)\|^2 = (u(\zeta_n) - u(\eta_n))^2 \frac{\pi}{2^{k+1}}. \qquad (5.30)$$

Since $|f'(t)| \le M_1$, then it satisfies the Lipschitz condition on each subinterval (i.e.,)

$$|u(\zeta_n) - u(\eta_n)| \le M_1|\zeta_n - \eta_n|, \; \forall \zeta_n, \eta_n \in \left[\frac{n}{2^k}, \frac{n+1}{2^k}\right]. \qquad (5.31)$$

Then from (5.30) and (5.31), we have

$$\|e_n(t)\|^2 \le \frac{\pi M_1^2}{2^{3k+1}}, \qquad (5.32)$$

which leads to

$$
\begin{aligned}
\|\tilde{e}(t)\|^2 &= \int_0^1 \tilde{e}(t)^2 w_n(t) dt = \int_0^1 \left(\sum_{n=1}^{2^k} e_n(t)\right)^2 w_n(t) dt \\
&= \int_0^1 \left(\sum_{n=1}^{2^k} e_n(t)^2\right) w_n(t) dt + 2\sum_{n \le n'} \int_0^1 e_n(t)e_{n'}(t)w_n(t) dt.
\end{aligned}
\qquad (5.33)
$$

Now due to the disjoint property of the basis functions, we have

$$\|\tilde{e}(t)\|^2 = \int_0^1 \left(\sum_{n=1}^{2^k} e_n(t)^2 \right) w_n(t) dt = \sum_{n=1}^{2^k} \|e_n(t)\|^2. \qquad (5.34)$$

By substituting (5.32) into (5.34) we obtain

$$\|\tilde{e}(t)\|^2 \le \frac{\pi M_1^2}{2^{2k+1}}, \qquad (5.35)$$

Or, in other words,

$$\|\tilde{e}(t)\| = O\left(\frac{1}{2^k} \right), \qquad (5.36)$$

which completes the proof.

When M is fixed, the larger the value of k, we get the more accurate the approximation solution.

5.4 Solving Linear Second-Order Two-Point Boundary Value Problems by S2KCWM

Consider the linear second-order differential equation

$$u''(x) + f_1(x)u'(x) + f_2(x)u(x) = G(x), \qquad x \in [0, 1], \qquad (5.37)$$

subject to the initial conditions

$$u(0) = \alpha, \quad u'(0) = \beta, \qquad (5.38)$$

(or) the boundary conditions

$$u(0) = \alpha, \quad u(1) = \beta, \qquad (5.39)$$

or the most general mixed boundary conditions

$$\alpha_1 u(0) + \alpha_2 u'(0) = \alpha, \quad b_1 u(1) + b_2 u'(1) = \beta. \qquad (5.40)$$

If we approximate the functions $u(x), f_1(x), f_2(x)$ and $G(x)$ regarding the second kind Chebyshev wavelet basis, one can write

$$u(x) \approx \sum_{n=0}^{2^k-1} \sum_{m=0}^{M} c_{nm} \psi_{nm}(x) = C^{\mathrm{T}} \psi(x). \quad f_1(x) \approx \sum_{n=0}^{2^k-1} \sum_{m=0}^{M} g_{nm} \psi_{nm}(x) = G_1^{\mathrm{T}} \psi(x)$$

$$(5.41)$$

$$f_2(x) \approx \sum_{n=0}^{2^k-1} \sum_{m=0}^{M} g_{nm} \psi_{nm}(x) = G_2^{\mathrm{T}} \psi(x) \quad G(x) = \sum_{n=0}^{2^k-1} \sum_{m=0}^{M} g_{nm} \psi_{nm}(x) = G^{\mathrm{T}} \psi(x)$$

$$(5.42)$$

Then

$$u'(x) \approx C^{\mathrm{T}} D \psi(x), \quad u''(x) = C^{\mathrm{T}} D^2 \psi(x) \tag{5.43}$$

Substitution of relations Eqs. (5.41), (5.42), and (5.43) into Eq. (5.37), enable us to define the residual, $R(x)$, of this equation, as

$$R(x) = C^{\mathrm{T}} D^2 \psi(x) + G_1^{\mathrm{T}} \psi(x)(\psi(x))^{\mathrm{T}} D^{\mathrm{T}} C + G_2^{\mathrm{T}} \psi(x)(\psi(x)^{\mathrm{T}} C - G^{\mathrm{T}} \psi(x). \tag{5.44}$$

and application of the Tau method yields the following $(2^k (M+1) - 2)$ linear equations in the unknown expansion coefficients, c_{nm}, namely

$$\int_0^t \sqrt{x - x^2} \psi_j(x) R(x) dx = 0, \quad j = 1, 2, \dots 2^k (M+1) - 2. \tag{5.45}$$

The initial conditions Eq. (5.38), the boundary conditions Eq. (5.39), and the mixed boundary conditions Eq. (5.40) lead, respectively, to the following equations

$$\begin{aligned} C^T \psi(0) = \alpha, \quad C^T D \psi(0) = \beta, \\ C^T \psi(0) = \alpha \quad C^T \psi(1) = \beta, \end{aligned} \tag{5.46}$$

and

$$a_1 C^T \psi(0) + a_2 D \psi(0) = \alpha, \quad b_1 C^T \psi(1) + b_2 C^T D \psi(1) = \beta. \tag{5.47}$$

Thus, Eq. (5.44) with the two equations of Eqs. (5.46) or (5.47) generate $2^k (M+1)$ of a set of linear equations. These equations can be solved for the unknown components of the vector C, and hence, the wavelet solution to $y(x)$ can be obtained.

5.4.1 Solving Nonlinear Second-Order Two-Point Boundary Value Problems by the S2KCWM

Consider the nonlinear differential equation

$$U''(x) = F(x, g(x), U'(x)), \tag{5.48}$$

Subject to the initial conditions

$$U(0) = \alpha, \quad U' = (0) = \beta, \tag{5.49}$$

or the boundary conditions

$$U(0) = \alpha, \quad U(1) = \beta, \tag{5.50}$$

or the most general mixed boundary conditions

$$\alpha_1 U(0) + \alpha_2 U'(0) = \alpha, \quad b_1 U(1) + b_2 U'(1) = \beta. \tag{5.51}$$

Using the scheme above, one can obtain

$$C^T D^2 \psi(t) = F(x, y(x), C^T D\psi(x)) \tag{5.52}$$

To find an approximate solution to $U(x)$, we compute Eq. (5.52) at the first $2^k(M + 1) - 2$ roots of $U^*_{2^k(M+1)}(x)$.

Equation (5.52) with Eqs. (5.51) or (5.50) or (5.49) generate $2^k(M + 1)$ nonlinear equations in the expansion coefficients C_{nm} which can be solved with the aid of Newton's iterative method.

5.5 Test Problems

Problem 5.1
We consider the nonlinear singular boundary value problem which arises in the study of the distribution of heat sources in the human head. This is also known as Emden–Fowler equation of the second kind.

$$\frac{d^2y}{dx^2} + \frac{2}{x}\frac{dy}{dx} = -e^{-y}, \tag{5.53}$$

and the boundary conditions

$$y'(0) = 0, \ y(1) + y'(1) = 0. \tag{5.54}$$

We solve Eq. (5.53) using the second kind shifting Chebyshev wavelet (S2KCW) algorithm described in Chap. 1 for the case corresponds to $M = 2$ and $k = 0$ to obtain the approximate solutions of $y(x)$.

The two operational matrices D and D^2 are given by

$$D = \begin{pmatrix} 0 & 0 & 0 \\ 4 & 0 & 0 \\ 0 & 8 & 0 \end{pmatrix}, \text{ and } D^2 = \begin{pmatrix} 0 & 0 & 0 \\ 0 & 0 & 0 \\ 32 & 0 & 0 \end{pmatrix}$$

Then, the function $\psi(r)$ can be obtained by using the following relation.

$$\psi(x) = \sqrt{\frac{2}{\pi}} \begin{pmatrix} 2 \\ 8x - 4 \\ 32x^2 - 32x + 6 \end{pmatrix}. \tag{5.55}$$

If we set

$$C = \left(c_{0,0}, c_{0,1}, c_{0,2}\right)^{\mathrm{T}} = \sqrt{\frac{\pi}{2}} (c_0, c_1, c_2)^{\mathrm{T}}, \tag{5.56}$$

then the S2KCW scheme is given by

$$C^{\mathrm{T}} D^2 \psi(x) + \frac{2}{x} C^{\mathrm{T}} D \psi(x) + \exp(-C^{\mathrm{T}} \psi(x)) = 0. \tag{5.57}$$

By selecting a root $x = \frac{2 - \sqrt{2}}{4}$, then we get the following equation

$$16c_1 - 35.8752c_2 + 0.14645 \left(e^{-(2c_0 - 2.828c_1 + 2c_2)}\right) = 0. \tag{5.58}$$

Using the boundary conditions, we gain

$$8c_1 - 32c_2 = 0, \tag{5.59}$$

$$2c_0 + 12c_1 + 38c_2 = 0. \tag{5.60}$$

On solving these Eqs. (5.58–5.60), we get

$$c_0 = 0.1578,$$
$$c_1 = -0.0147,$$
$$c_2 = -0.0037.$$

Table 5.1 Approximate solutions for Problem 5.1

x	S2KCWM	Theory of maximum principles [2]	Error
0.1000	0.3510	0.3664	0.0154
0.2000	0.3475	0.3629	0.0154
0.3000	0.3415	0.3571	0.0156
0.4000	0.3333	0.3489	0.0156
0.5000	0.3226	0.3384	0.0158
0.6000	0.3096	0.3254	0.0158
0.7000	0.2942	0.3010	0.0068
0.8000	0.2764	0.2920	0.0156
0.9000	0.2562	0.2713	0.0151
1.0000	0.2238	0.2479	0.0241

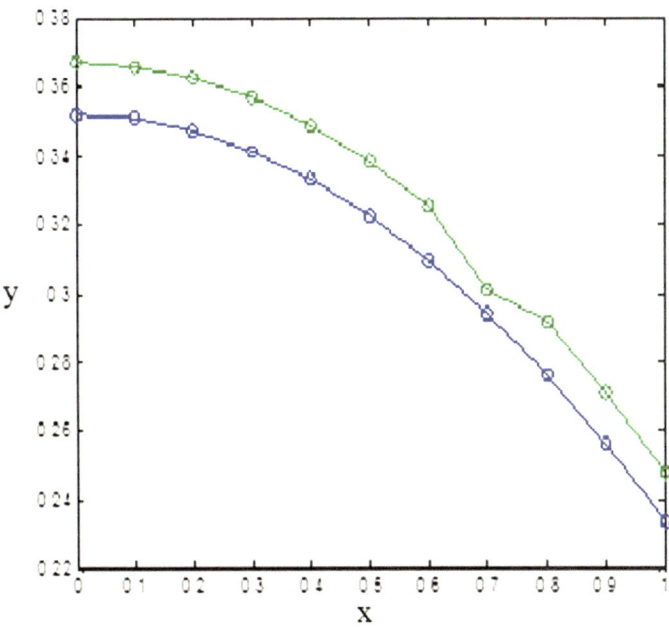

Fig. 5.1 Comparison of solutions of S2KCWM ($M = 2$, $k = 0$) (lower curve) and solutions of theory of maximum principles (upper curve) for Problem 5.1

Then (Table 5.1 and Fig. 5.1).

$$y(x) = -0.1184x^2 + 0.3522. \qquad (5.61)$$

Problem 5.2

We consider the above nonlinear singular boundary value problem (heat conduction model of the human head) with different boundary conditions

$$\frac{d^2y}{dx^2} + \frac{2}{x}\frac{dy}{dx} = -e^{-y}, \tag{5.62}$$

and the boundary conditions

$$y'(0) = 0, \, 2y(1) + y'(1) = 0, \tag{5.63}$$

then the SK2CW scheme is given by

$$C^T D^2 \psi(x) + \frac{2}{x} C^T D\psi(x) + \exp(-C^T \psi(x)) = 0. \tag{5.64}$$

If we select a root $x = \frac{2-\sqrt{2}}{4}$, then we get the following equation

$$16c_1 - 35.8752c_2 + 0.14645\left(e^{-(2c_0 - 2.828c_1 + 2c_2)}\right) = 0, \tag{5.65}$$

and using the boundary conditions, we gain

$$8c_1 - 32c_2 = 0, \tag{5.66}$$

$$4c_0 + 16c_1 + 44c_2 = 0. \tag{5.67}$$

On solving these Eqs. (5.65–5.67), we get

$$c_0 = 0.1089,$$
$$c_1 = -0.01613,$$
$$c_2 = -0.004034.$$

Consequently, (Table 5.2 and Fig. 5.2).

$$y(x) = -0.1291x^2 + 0.2799. \tag{5.68}$$

Problem 5.3

Consider another singular boundary value problem arising in Physiology [4]

$$\frac{d^2y}{dx^2} + \frac{1}{x}\frac{dy}{dx} = -e^y, \tag{5.69}$$

and the boundary conditions

Table 5.2 Approximate solutions for Problem 5.2

x	S2KCWM	Theory of maximum principles [2]	Error
0.1000	0.2786	0.2691	0.0095
0.2000	0.2747	0.2653	0.0094
0.3000	0.2683	0.2589	0.0094
0.4000	0.2592	0.2499	0.0093
0.5000	0.2476	0.2382	0.0094
0.6000	0.2334	0.2239	0.0095
0.7000	0.2166	0.2068	0.0098
0.8000	0.1973	0.1868	0.0105
0.9000	0.1753	0.1639	0.0114
1.0000	0.1508	0.1379	0.0129

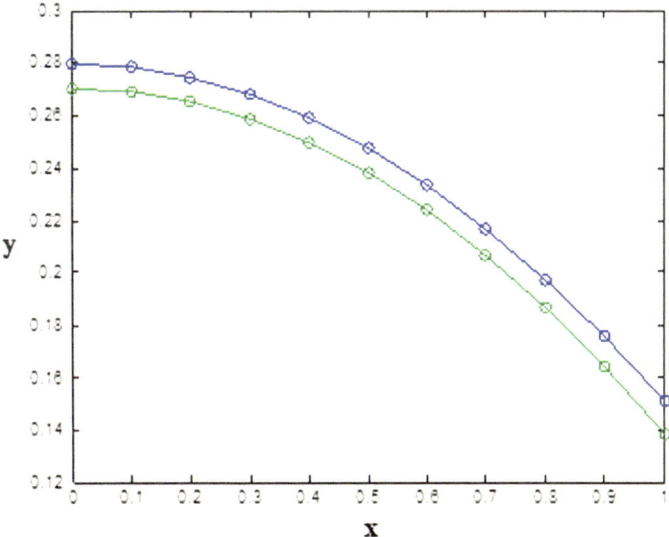

Fig. 5.2 Comparison of solutions of S2KCWM ($M = 2$, $k = 0$) (upper curve) and solutions of theory of maximum principles (lower curve) for Problem 5.2

$$y'(0) = 0, y(1) = 0. \tag{5.70}$$

The exact solution is $y(x) = 2\ln\left(\frac{c+1}{cx^2+1}\right)$,

where $c = 3 - 2\sqrt{2}$

then the shifted S2KCW scheme is given by

$$C^\mathrm{T} D^2 \psi(x) + \frac{1}{x} C^\mathrm{T} D\psi(x) + \exp(C^\mathrm{T}\psi(x)) = 0. \tag{5.71}$$

If we select a root $x = \frac{2-\sqrt{2}}{4}$, then we get the following equation

$$8c_1 - 13.2512c_2 + 0.14645\left(e^{(2c_0 - 2.828c_1 + 2c_2)}\right) = 0, \tag{5.72}$$

And using the boundary conditions, we gain

$$8c_1 - 32c_2 = 0, \tag{5.73}$$

$$2c_0 + 4c_1 + 6c_2 = 0. \tag{5.74}$$

On solving these Eqs. (5.72–5.74), we get

$$c_0 = 0.1214,$$
$$c_1 = -0.0441,$$
$$c_2 = -0.0110.$$

Now (Table 5.3 and Fig. 5.3).

$$y(x) = -0.353x^2 + 0.353. \tag{5.75}$$

Problem 5.4
We consider another singular boundary value problem which arises in the study of steady-state oxygen diffusion in a spherical cell [5].

$$\frac{d^2 y}{dx^2} + \frac{2}{x}\frac{dy}{dx} = \frac{0.76129y}{y + 0.03119}, \tag{5.76}$$

x	Exact	S2KCWM $M = 2, k = 0$	Absolute numerical error
0.1000	0.3133	0.3495	0.0362
0.2000	0.3031	0.3389	0.0358
0.3000	0.2861	0.3212	0.0351
0.4000	0.2626	0.2965	0.0339
0.5000	0.2327	0.2647	0.0320
0.6000	0.1969	0.2259	0.0290
0.7000	0.1553	0.1800	0.0247
0.8000	0.1083	0.1271	0.0188
0.9000	0.0564	0.0671	0.0107
1.0000	0.0000	0.0000	0.0000

Table 5.3 Numerical errors for Problem 5.3

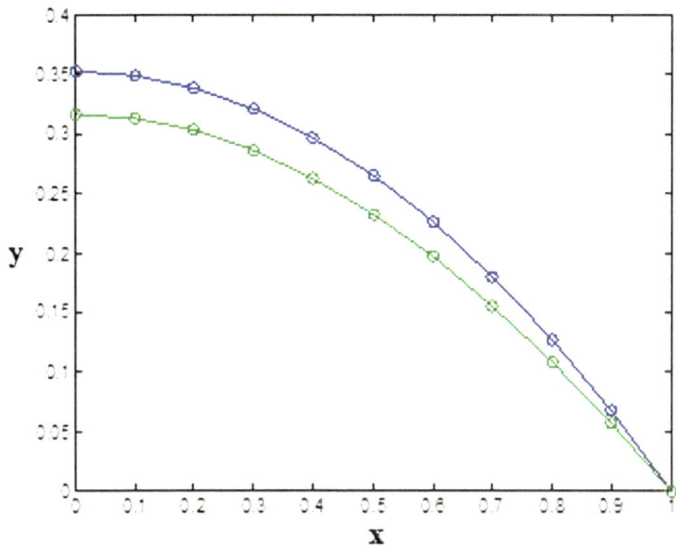

Fig. 5.3 Comparison of exact (lower curve) and S2KCWM (upper curve) solutions for Problem 5.3

and the boundary conditions

$$y'(0) = 0, \quad 5y(1) + y'(1) = 5, \tag{5.77}$$

then the shifted S2KCW scheme is given by

$$C^T D^2 \psi(x) + \frac{2}{x} C^T D\psi(x) - \frac{0.76129 C^T \psi(x)}{C^T \psi(x) + 0.03119} = 0. \tag{5.78}$$

If we select a root $x = \frac{2-\sqrt{2}}{4}$, then we get

$$\begin{aligned}
&- 0.223c_0 + 0.8143c_1 - 1.342c_2 - 45.248c_1^2 \\
&- 71.750c_2^2 + 32c_0c_1 - 71.750c_0c_2 + 133.4553c_1c_2 = 0
\end{aligned} \tag{5.79}$$

and using the boundary conditions, we gain

$$8c_1 - 32c_2 = 0, \tag{5.80}$$

$$10c_0 + 28c_1 + 62c_2 = 0. \tag{5.81}$$

On solving these Eqs. (5.79–5.81), we gain

$$c_0 = 0.4335, \quad c_0 = 0.09076,$$
$$c_1 = 0.0153, \quad c_1 = 0.09408,$$
$$c_2 = 0.0038. \quad c_2 = 0.02352.$$

Now we get two solutions, respectively, for the two set of values which satisfy boundary conditions

$$y(x) = 0.1223x^2 + 0.8288. \tag{5.82}$$

$$y(x) = 0.75264x^2 - 0.05368. \tag{5.83}$$

From the figure we can conclude that Eq. (5.82) is the suitable solution for Problem 5.4 (Figs. 5.4, 5.5 and Table 5.4).

Problem 5.5

Consider the nonlinear singular boundary value problem describing the equilibrium of isothermal gas sphere [6]

$$\frac{d^2y}{dx^2} + \frac{2}{x}\frac{dy}{dx} + y^5 = 0, 0 < x < 1, \tag{5.84}$$

and the boundary conditions

$$y'(0) = 0, y(0) = 1. \tag{5.85}$$

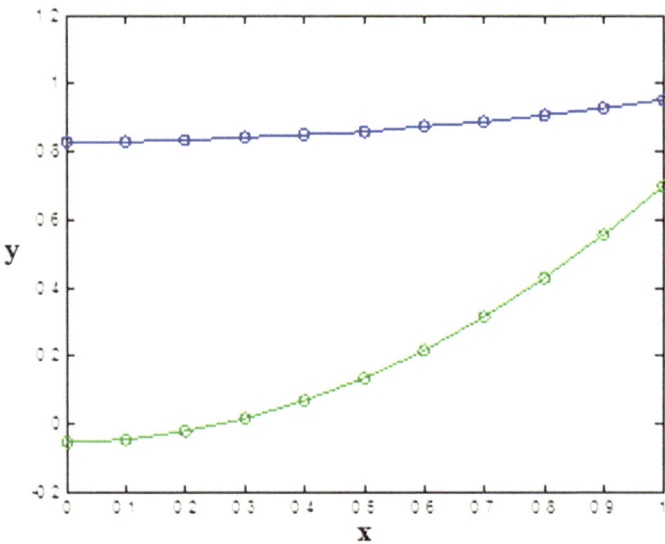

Fig. 5.4 Comparison of Eq. (5.82) (upper curve) and Eq. (5.83) (lower curve)

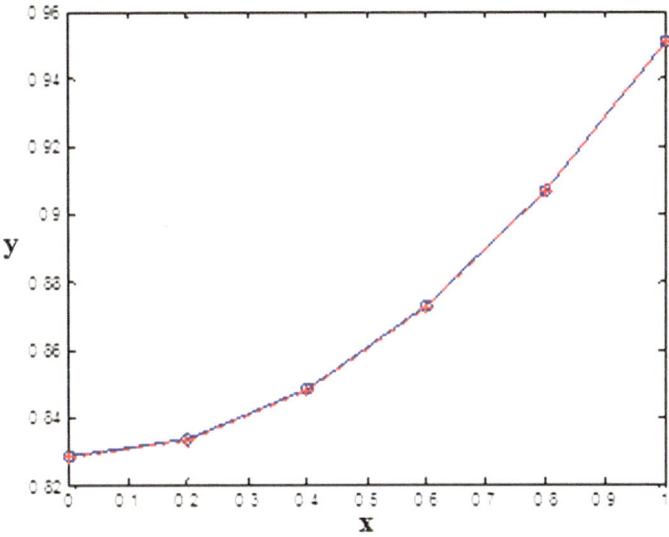

Fig. 5.5 Comparison of S2KCWM ($M = 2$, $k = 0$) (o- curve) and VIM (+- curve) solutions for Problem 5.4

Table 5.4 Approximate solutions for Problem 5.4

x	S2KCWM $M = 2$, $k = 0$	S2KCWM $M = 3$, $k = 1$	VIM
0.2	0.8337	0.8336	0.8334
0.4	0.8484	0,8483	0.8481
0.6	0.8728	0.8726	0.8725
0.8	0.9071	0.9070	0.9068
1.0	0.9511	0.9511	0.9511

The exact solution is $y(x) = \sqrt{\frac{3}{3+x^2}}$.

This is also known as Emden–Fowler equation of the first kind.

Then the shifted S2KCW scheme is given by

$$C^T D^2 \psi(x) + \frac{2}{x} C^T D\psi(x) + (C^T \psi(x))^5 = 0. \tag{5.86}$$

If we select a root $x = \frac{2-\sqrt{2}}{4}$, then we get the following equation

$$16c_1 - 35.8752c_2 + 0.14645(2c_0 - 2.828c_1 + 2c_2)^5 = 0, \tag{5.87}$$

and using the boundary conditions, we gain

Table 5.5 Numerical errors for Problem 5.5

x	Exact	S2KCWM ($M = 2$, $k = 0$)	Absolute numerical error
0.2	0.9934	0.9933	0.0001
0.4	0.9744	0.9736	0.0008
0.6	0.9449	0.9408	0.0041
0.8	0.9078	0.8949	0.0129
1.0	0.8660	0.8358	0.0302

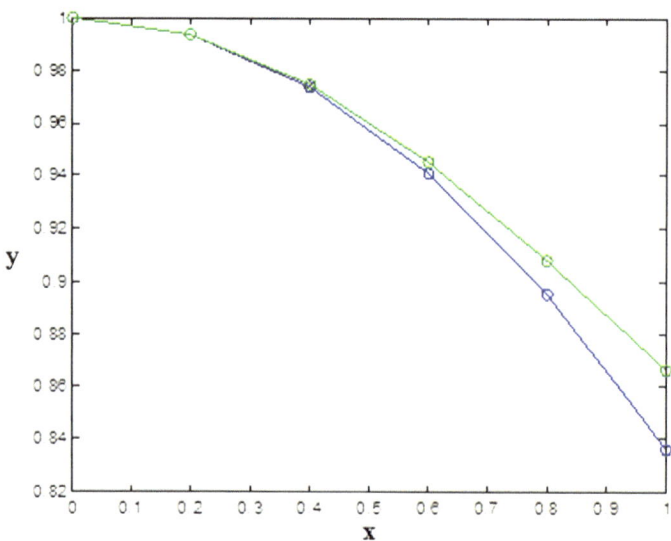

Fig. 5.6 Comparison of exact (upper curve) and S2KCWM ($M = 2$, $k = 0$) (lower curve) solutions for Problem 5.5

$$8c_1 - 32c_2 = 0, \tag{5.88}$$

$$2c_0 - 4c_1 + 6c_2 = 1. \tag{5.89}$$

On solving these Eqs. (5.87–5.89), we get

$$c_0 = 0.47442,$$
$$c_1 = -0.02050,$$
$$c_2 = -0.00512.$$

Other four sets of solutions are complex numbers.
Then (Table 5.5 and Fig. 5.6).

$$y(x) = 0.164x^2 - 0.99984.$$ (5.90)

Problem 5.6

Consider another nonlinear singular boundary value problem which arises in the radial stress on a rotationally symmetric shallow membrane cap [6]

$$\frac{d^2y}{dx^2} + \frac{3}{x}\frac{dy}{dx} = \frac{1}{2} - \frac{1}{8y^2(x)}, 0 \le x \le 1,$$ (5.91)

and the boundary conditions

$$y'(0) = 0, y(1) = 1.$$ (5.92)

Then the shifted S2KCW scheme is given by

$$C^T D^2 \psi(x) + \frac{3}{x} C^T D \psi(x) - \frac{1}{2} + \frac{1}{8(C^T\psi(x))^2} = 0.$$ (5.93)

If we select a root $x = \frac{2-\sqrt{2}}{4}$, then we get the following equations

$$512c_2(2c_0 - 2.828c_1 + 2c_2)^2 + 163.8785(2c_0 - 2.828c_1 + 2c_2)^2$$
$$\times (8c_1 - 22.624c_2) - 4(2c_0 - 2.828c_1 + 2c_2)^2 = -1,$$ (5.94)

and using the boundary conditions, we gain

$$8c_1 - 32c_2 = 0,$$ (5.95)

$$2c_0 + 4c_1 + 6c_2 = 1.$$ (5.96)

On solving these Eqs. (5.94–5.96), we get

$$c_0 = 0.484400,$$
$$c_1 = 0.005672,$$
$$c_2 = 0.001418.$$

Other two sets of solutions are complex numbers.
Then (Table 5.6 and Fig. 5.7).

$$y(x) = 0.045376x^2 + 0.95462.$$ (5.97)

Table 5.6 Approximate solutions for Problem 5.6

x	S2KCWM ($M = 2$, $k = 0$)	ADM	VIM
0.1	0.9551	0.9546	0.9526
0.2	0.9564	0.9559	0.9541
0.3	0.9587	0.9582	0.9565
0.4	0.9619	0.9614	0.9599
0.5	0.9660	0.9655	0.9642
0.6	0.9710	0.9705	0.9695
0.7	0.9769	0.9765	0.9757
0.8	0.9837	0.9834	0.9829
0.9	0.9914	0.9912	0.9910
1.0	1.0000	1.0000	1.0000

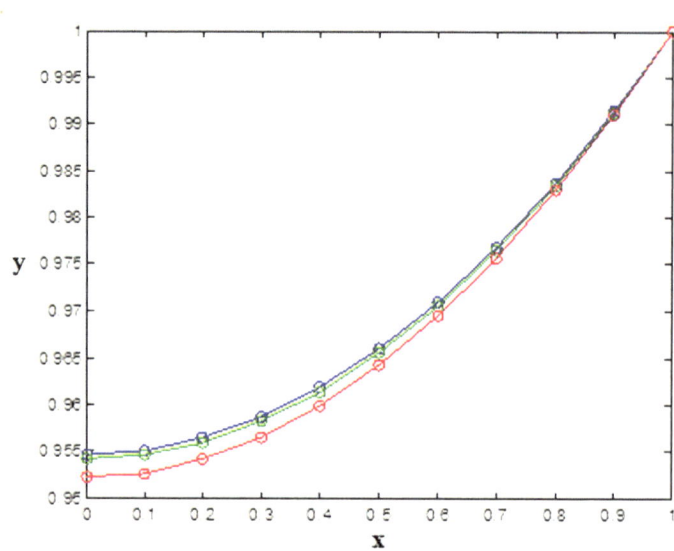

Fig. 5.7 Comparison among solutions of ADM (middle curve) and VIM (lower curve) and S2KCWM ($M = 2$, $k = 0$) (upper curve) for Problem 5.6

Problem 5.7

Consider the following nonlinear fractional spherical isothermal Lane–Emden-type equation

$$D^{\alpha}y(x) + \frac{2}{x^{\alpha-\beta}} D^{\beta}y(x) = e^{-y(x)}, 0 < x \leq 1, \tag{5.98}$$

Subject to the boundary conditions

$$y(0) = 1, y'(0) = 0. \tag{5.99}$$

where $y = (x)$ is the unknown function. $D^\alpha y(x)$ and $D^\beta y(x)$ are the modified Riemann–Liouville derivatives.

At $\alpha = \frac{3}{2}, \beta = \frac{3}{4}$, the power series solution is given by

$$y(x) = 0.2368456765x^{\frac{3}{2}} - 0.02568610429x^3 + 0.005004915254x^{\frac{9}{2}}$$
$$- 0.00129847099x^6 + 0.0004026307151x^{\frac{15}{2}} - \dots \tag{5.100}$$

The fractional derivative through fractional difference is given as follows:

Let $f : R \to R, x \to f(x)$ denote the continuous function and let h indicates the constant discretization span, the limit form of the modified Riemann–Liouville derivative is given by

$$f^{(\alpha)}(x) = \lim_{h \to 0} \frac{\Delta^\alpha[f(x) - f(0)]}{h^\alpha}, 0 < \alpha < 1, \tag{5.101}$$

where

$$\Delta^\alpha f(x) = \sum_{k=0}^{\infty} (-1)^k \frac{\Gamma(\alpha+1)}{\Gamma(k+1)\Gamma(\alpha-k+1)} f[x + (\alpha-k)h], \tag{5.102}$$

and α-order derivative of a constant is zero.

The integral form of modified Riemann–Liouville derivative is given as

$$D_x^\alpha f(x) = \begin{cases} \frac{1}{\Gamma(-\alpha)} \int\limits_0^x (x-\xi)^{-\alpha-1}[f(\xi) - f(0)]d\xi, & \alpha < 0 \\ \frac{1}{\Gamma(1-\alpha)} \frac{d}{dx} \int\limits_0^x (x-\xi)^{-\alpha}[f(\xi) - f(0)]d\xi, & 0 < \alpha < 1 \\ \frac{1}{\Gamma(n-\alpha)} \frac{d^n}{dx^n} \int\limits_0^x (x-\xi)^{n-\alpha-1}[f(\xi) - f(0)]d\xi, & n \le \alpha < n+1, n \ge 1 \end{cases}$$
$$\tag{5.103}$$

Following properties of modified Riemann–Liouville derivative hold:

$$D_x^\alpha x^\gamma = \frac{\Gamma(\gamma+1)}{\Gamma(\gamma+1-\alpha)} x^{\gamma-\alpha}, \quad \gamma > 0. \tag{5.104}$$

$$D_x^\alpha (cf(x)) = cD_x^\alpha f(x), \tag{5.105}$$

where c is a constant.

$$D_x^\alpha f(x) \cong \Gamma(\alpha+1)D_x f(x). \tag{5.106}$$

From the above result, the following equalities hold, which are

$$D_x^\alpha[f(x)g(x)] = g(x)D_x^\alpha f(x) + f(x)D_x^\alpha g(x). \tag{5.107}$$

$$D_x^\alpha f(x) \cong \Gamma(\alpha+1)Df(x). \tag{5.108}$$

$$D_x^\alpha f[g(x)] = f_g[g(x)]D_x^\alpha g(x). \tag{5.109}$$

The S2KCW matrix is

$$\psi(x) = \sqrt{\frac{2}{\pi}} \begin{pmatrix} 2 \\ 8x - 4 \\ 32x^2 - 32x + 6 \end{pmatrix} \tag{5.110}$$

Using modified Riemann–Liouville derivative, we have

$$D^{\frac{1}{2}}\psi(t) = \begin{pmatrix} 0 \\ 9.0270t^{\frac{1}{2}} \\ 48.1442t^{\frac{3}{2}} - 36.1081t^{\frac{1}{2}} \end{pmatrix} \quad D^{\frac{1}{3}}\psi(t) = \begin{pmatrix} 0 \\ 8.8624t^{\frac{2}{3}} \\ 42.5362t^{\frac{5}{3}} - 35.4496t^{\frac{2}{3}} \end{pmatrix}$$

$$D^{\frac{3}{2}}\psi(t) = \begin{pmatrix} 0 \\ 4.5136t^{-\frac{1}{2}} \\ 72.2185t^{\frac{1}{2}} - 18.0544t^{-\frac{1}{2}} \end{pmatrix} \quad D^{\frac{12}{10}}\psi(t) = \begin{pmatrix} 0 \\ 6.8717t^{-\frac{2}{10}} \\ 68.7152t^{\frac{8}{10}} - 27.4867t^{-\frac{2}{10}} \end{pmatrix} \tag{5.111}$$

Here $D^{\frac{3}{2}}$ and $D^{\frac{3}{4}}$ are fractional differential operators.
If we set

$$C = \left(c_{0,0}, c_{0,1}, c_{0,2}\right)^{\mathrm{T}} = \sqrt{\frac{\pi}{2}}(c_0, c_1, c_2)^{\mathrm{T}}, \tag{5.112}$$

then the S2KCW scheme without using the operational matrices of derivatives is given by

$$C^{\mathrm{T}}D^{\frac{3}{2}}\psi(x) + \frac{2}{x^{\frac{3}{4}}}C^{\mathrm{T}}D^{\frac{3}{4}}\psi(x) + \exp(-C^{\mathrm{T}}\psi(x)) = 0. \tag{5.113}$$

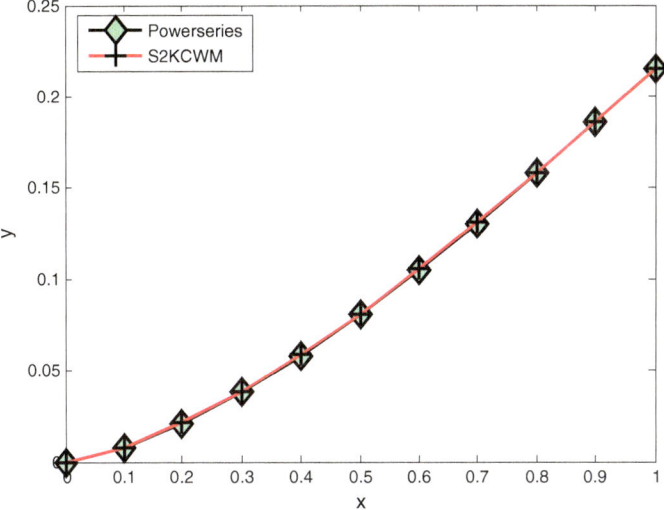

Fig. 5.8 Comparison between the power series solution and S2KCWM solution for Problem 5.7

Using boundary conditions, we gain

$$C^T \psi(0) = 1, \tag{5.114}$$

$$C^T \psi'(0) = 0. \tag{5.115}$$

On collocating with $x = \frac{2-\sqrt{2}}{4}$, we get a set of three algebraic equations. On solving, we get the solutions given in Fig. 5.8.

Problem 5.8
Consider the following fractional Lane–Emden-type equation

$$D^\alpha y(x) + \frac{k}{x^{\alpha-\beta}} D^\beta y(x) + \frac{1}{x^{\alpha-2}} y(x) = h(x), 0 < x \le 1, \tag{5.116}$$

Subject to the boundary conditions

$$y(0) = 0, y'(0) = 0, \tag{5.117}$$

where

$$h(t) = t^{2-\alpha} \left(-6t \left(\frac{t^2}{6} + \frac{\Gamma 4 - \beta + k\Gamma 4 - \alpha}{\Gamma 4 - \beta \Gamma 4 - \alpha} \right) + 2 \left(\frac{t^2}{2} + \frac{\Gamma 3 - \beta + k\Gamma 3 - \alpha}{\Gamma 3 - \beta \Gamma 3 - \alpha} \right) \right). \tag{5.118}$$

At $\alpha = \frac{3}{2}, \beta = 1$, the exact solution is

$$y = t^2 - t^3. \tag{5.119}$$

The S2KCW matrices are given by

$$\psi(x) = \sqrt{\frac{2}{\pi}} \begin{pmatrix} 2 \\ 8x - 4 \\ 32x^2 - 32x + 6 \end{pmatrix}, \quad D^{\frac{3}{2}}\psi(x) = \begin{pmatrix} 0 \\ 4.5134x^{-\frac{1}{2}} \\ 72.2163x^{\frac{1}{2}} - 18.0541x^{-\frac{1}{2}} \end{pmatrix} \tag{5.120}$$

Here D is ordinary differential operator and $D^{\frac{3}{2}}$ is fractional differential operator. Then the S2KCW scheme without using the operational matrices of derivatives is given by

$$C^T D^{\frac{3}{2}}\psi(x) + \frac{2}{x^{\frac{3}{2}}} C^T D\psi(x) + x^{\frac{1}{2}} C^T \psi(x) = 0. \tag{5.121}$$

Using boundary conditions, we gain

$$C^T \psi(0) = 0, \tag{5.122}$$

$$C^T \psi'(0) = 0. \tag{5.123}$$

On collocating with $x = \frac{2-\sqrt{2}}{4}$, we get a set of three algebraic equations. On solving them, we get the solutions given in Table 5.7 and Fig. 5.9.

Table 5.7 Comparison between exact and S2KCWM solutions for Problem 5.8

x	Exact	S2KCWM ($M = 2$, $k = 0$)	Absolute error
0.1	0.0090	0.0096	6×10^{-4}
0.2	0.0320	0.0325	5×10^{-4}
0.3	0.0630	0.0635	5×10^{-4}
0.4	0.0960	0.0964	4×10^{-4}
0.5	0.1250	0.1254	4×10^{-4}
0.6	0.1440	0.1443	3×10^{-4}
0.7	0.1470	0.1472	2×10^{-4}
0.8	0.1280	0.1284	4×10^{-4}
0.9	0.0810	0.0816	6×10^{-4}
1.0	0.0000	0.0000	0.0000

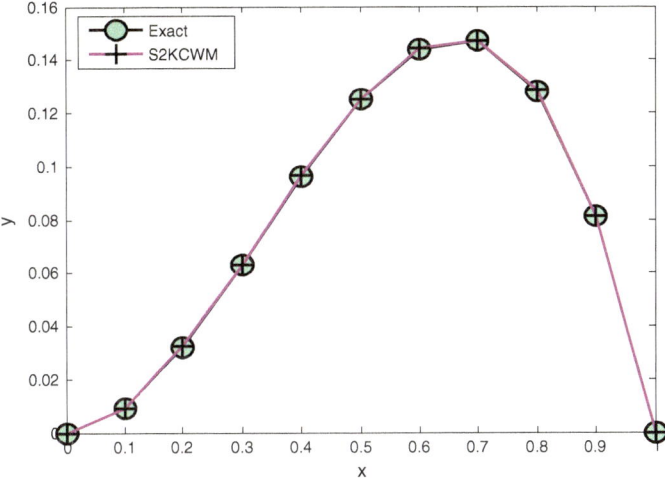

Fig. 5.9 Comparison between the exact solution and S2KCWM solution for the Problem 5.8

	x	Method	Error	k, M	Runtime (in seconds)
Table 5.8 Numerical experiments are presented in double precision with some MATLAB codes on a computer system Intel(R) Core (TM) i54210U CPU@2.40 GHz	0.1	S2KCWM	2.610E−9	0,2	1.97
		S2KCWM	2.112E−9	1,3	1.88
		VIM	4.501E−9	...	62.36
	0.5	S2KCWM	2.401E−9	0,2	2.73
		S2KCWM	2.005E−9	1,3	2.61
		VIM	3.721E−9	...	58.23

5.6   Comparison of Maximum Error with Computation Run Time (in Seconds) for Numerical Experiment

Table 5.8

5.7   Conclusion

In this chapter, an efficient Chebyshev wavelet-based approximation algorithm is developed for solving nonlinear singular boundary value problems. The proposed results show that the S2KCWM can match the analytical solution very efficiently. Also, the proposed method has a simple implementation process. The advantage of this algorithm is that highly accurate approximate solutions are achieved using a very few number of terms of the estimated expansion. The proposed wavelet scheme can overcome the difficulty of singularity in nonlinear boundary value problems easily. Also, the calculations involved in S2KCWM are simple, straightforward, and less computation cost.

References

1. S. Lin, Oxygen diffusion in s spherical cell with nonlinear oxygen uptake kinetics. J. Theor. Biol. **60**(2), 449–457 (1976)
2. R. Duggan, A. Goodman, Pointwise bounds for a nonlinear heat conduction model of the human head. Bull. Math. Biol. **48**(2), 229–236 (1986)
3. R.K. Pandey, A.K. Singh, On the convergence of finite difference method for a class of singular boundary value problems arising in physiology. J. Comput. Appl. Math. **166**, 553–564 (2004)
4. S. Khuri, A. Safy, A novel approach for the solution of a class of singular boundary values problems arising in physiology. Math. Comput. Model. **52**(3), 626–636 (2010)
5. A. Wazwaz, The variational iteration method for solving nonlinear singular boundary problems arising in various physical models. Commun. Nonlinear Sci. Numer. Simul. **16**(10), 3881–3886 (2011)
6. A. Ravikanth, K. Aruna, He's variational iteration method for treating nonlinear singular boundary value problems. Comput. Math Appl. **60**(3), 821–829 (2010)
7. R. Singh, J. Kumar, An efficient numerical technique for the solution of nonlinear singular boundary value problems. Comput. Phys. Commun. (2014) (in press)
8. H. Caglar, N. Caglar, M. Ozer, B-spline solution of non linear singular boundary value problems arising in physiology. Chaos, Solitons Fractals **39**, 1232–1237 (2009)
9. R. Singh, J. Kumar, Solving a class of singular two-point boundary value problems using new modified decomposition method. ISRN Comput. Math. **2013**, 1–11 (2013)
10. Z. Cen, Numerical method for a class of singular non linear boundary value problems using green's functions. Int. J. C. Math. **24**(3–4), 29–310 (1988)
11. M. Inc, M. Engut, Y. Cherrualt (2005) A different approach for solving singular two-point boundary value problems. kybernetes. **34**(7), 934–940 (The international journal of system and Cybernetics)
12. H. Adibi, P. Assari, Chebyshev wavelet method for numerical solution of Fredholm integral equations of the first kind. Math. Probl. Eng. (2010) Article ID 138408
13. Y. Li, Solving a nonlinear fractional differential equation using Chebyshev wavelets. Commun. Nonlinear Sci. Numer. Simulat. **11**, 2284–2292 (2011)
14. S. Sohrabi, Comparison Chebyshev wavelet method with BPFs method for solving Abel's integral equation. Ain Shams Eng. J. **2**, 249–254 (2011)
15. J.C. Mason, C. David, *Handscomb* (Taylor and Francis, Chebyshev polynomials, 2002)
16. M. Ghasemi, M.T. Kajani, Numerical solution of time-varying delay systems by Chebyshev wavelets. Appl. Math. Model. **35**, 5235–5244 (2011)
17. W.M. Abd-Elhameed, E.H. Doha, Y.H. Youssri, New wavelets collocation method for solving second-order multipoint boundary value problems using Chebyshev polynomials of third and fourth kinds. Abstr. Appl. Anal. (2013) Article ID 542839
18. W. M. Abd-Elhameed, E. H. Doha, Y. H. Youssri, New spectral second kind Chebyshev Wavelets algorithm for solving linear and nonlinear second-order differential equations involving singular and Bratu type equations. Abstr. Appl. Anal. (2013) Article ID 715756
19. SGh Hosseini, A new operational matrix of derivative for Chebyshev wavelets and its applications in solving ordinary differential equations with non analytic solution. Appl. Math. Sci. **5**(51), 2537–2548 (2011)

20. A. Barzkar, M.K. Oshagh, P. Assari, M.A. Mehrpouya, Numerical solution of the nonlinear Fredholm integral equation and the Fredholm integro-differential equation of second kind using Chebyshev wavelets. World Appl. Sci. J. **18**(12), 1774–1782 (2012)
21. E. Babolian, F. Fattahzadeh, Numerical solution of differential equations by using Chebyshev wavelet operational matrix of integration. Appl. Math. Comput. **188**(1), 417–426 (2007)
22. L. Zhu, Q. Fan, Solving fractional nonlinear Fredholm integro-differential equations by the second kind Chebyshev wavelet. Commun. Nonlinear Sci. Numer. Simul. **17**(6), 2333–2341 (2012)
23. E.H. Doha, W.M. Abd- Elhameed, Y.H. Youssri, Second kind Chebyshev operational matrix algorithm for solving differential equations of Lane-Emden type. New Astron. **23–24**, 113–117 (2013)
24. E.H. Doha, W.M. Abd-Elhameed, M.A. Bassuony, New algorithms for solving high even-order differential equations using third and fourth Chebyshev-Galerkin methods. J. Comput. Phys. **236**, 563–579 (2013)
25. G. Hariharan, An efficient wavelet based approximation method to water quality assessment model in a uniform channel. Ains. Shams. Eng. J. (2013) (in Press)
26. G. Hariharan, K. Kannan, K.R. Sharma, Haar wavelet in estimating the depth profile of soil temperature. Appl. Math. Comput. **210**, 119–225 (2009)
27. G. Hariharan, K. Kannan, Haar wavelet method for solving Fisher's equation. Appl. Math. Comput. **211**, 284–292 (2009)
28. G. Hariharan, K. Kannan, A Comparative Study of a Haar Wavelet method and a restrictive Taylor's series method for solving convection-diffusion equations. Int. J. Comput. Methods Eng. Sci. Mech. **11**(4), 173–184 (2010)
29. G. Hariharan, K. Kannan, Review of wavelet methods for the solution of reaction–diffusion problems in science and engineering. Appl. Math. Model. **38**(1), 799–813 (2014)

Chapter 6
Analytical Expressions of Amperometric Enzyme Kinetics Pertaining to the Substrate Concentration Using Wavelets

In this chapter, wavelet-based operational matrix methods have been developed to investigate the approximate solutions for amperometric enzyme kinetic problems. The operational matrices of derivatives have been utilized for solving the nonlinear initial value problems. The accuracy of the proposed wavelet-based approximation methods has been confirmed. The main purpose of the proposed method is to get better and more accurate results. Operational matrices of Chebyshev and Legendre wavelets are utilized to obtain a sequence of discrete equations into the systems of algebraic equations and the solutions of algebraic systems lead to the solution of nonlinear initial value problems. Numerical experiments are given to demonstrate the accuracy and efficiency of the proposed method.

6.1 Introduction

Amperometric biosensors are sensors which measure the response of an electrode due to direct electrochemical oxidation or reduction of the products of a biocatalytic reaction. In biocatalyst reactions, enzyme acts as a catalyst for chemical transformations of compounds. The response of amperometric biosensors is determined by the enzyme activity. In recent years, several mathematical models have been proposed to study the analytical characteristics of amperometric enzyme kinetics [1]. Bartlett and Whitaker [2] derived the analytic expressions related to the response of amperometric enzyme electrode. Baronas et al. [3] designed a mathematical model for cyclic substrate conversion reactions taking place at a single enzyme membrane. Baronas and Kulys [4] developed a theoretical model of an amperometric enzyme electrode with substrate cyclic conversion.

In general, it is not easy to derive the analytical solutions to most of the nonlinear differential equations. Therefore, it is vital to develop some reliable and efficient techniques to solve nonlinear differential equations. In recent years, numerous applications of ordinary and partial differential equations have appeared in many

© Springer Nature Singapore Pte Ltd. 2019   93
G. Hariharan, *Wavelet Solutions for Reaction–Diffusion Problems in Science and Engineering*, Forum for Interdisciplinary Mathematics,
https://doi.org/10.1007/978-981-32-9960-3_6

areas of physics and engineering. It is well known that the numerical methods have played a significant role in solving nonlinear differential equations. Therefore, many numerical methods have been developed and applied for providing approximate solutions. He and his co-workers [5, 9] introduced the homotopy perturbation method (HPM) for the nonlinear differential equations. He and Wu [6] introduced a novel variational iteration method (VIM) for a few nonlinear differential equations. Malvandi and Ganji [7] applied the homotopy perturbation method (HPM) and VIM for the approximate solutions for amperometric enzyme kinetic problems. They concluded that HPM provides better convergence for nonlinear differential equations arising in engineering [4, 6, 8]. He [2, 9] proposed a novel method for the solutions of nonlinear differential equations. Rajendran et al. [5] established the VIM for the numerical solutions for Michaelis–Menten enzyme kinetic problems. Malvandi and Ganji [7] used Pade's approximation method for the enzyme kinetic-related problems. Rahamathunissa and Rajendran [10] used He's variational iteration method for the numerical solutions of nonlinear boundary value problems arising in enzyme–substrate reaction–diffusion processes. Meena and Rajendran [11] developed the perturbation method for the approximate/analytical expressions for substrate and product concentration, current response for all values of parameters. Eswari and Rajendran [12] obtained the HPM solutions for the analytical expressions of concentration and current in homogeneous catalytic reactions at spherical microelectrodes. Indira and Rajendran [1] developed the analytical expressions of the concentration and substrates and product in phenol–polyphenol oxidase system immobilized in laponite hydrogels Michaelis–Menten formalism in homogeneous medium. Indira and Rajendran [1] discussed a mathematical modeling in amperometric oxidase enzyme–membrane electrodes. A semi-analytical method has been developed for the approximate/analytical expressions for concentrations in amperometric oxidase enzyme electrode at steady-state conditions. Shanmugarajan et al. [13] derived the analytical expressions for the electrochemical polymerizations. Rajendran and Anitha [5] established the homotopy perturbation method (HPM) for the analytical expressions for amperometric enzymatic reaction–diffusion equations. Malvandi and Ganji [7] obtained the analytical solutions by the Pade approximation of rational expressions for enzyme kinetic problems. Dohaa et al. [14] established the highly accurate approximate solutions for a few number of second kind Chebyshev wavelets. Recently, Sathyaseelan et al. [15] introduced wavelet approximation method for a few nonlinear oscillator equations.

In this chapter, two reliable Chebyshev and Legendre wavelet methods have been implemented for the approximate/analytical solutions for amperometric enzyme kinetic problems. To the best of our knowledge, until now there is no operational matrix method has been addressed for the above enzyme kinetic problems.

This chapter is organized as follows. In Sect. 6.2, mathematical formulation of the amperometric enzyme kinetics is discussed. Legendre and Chebyshev wavelets are described in Sect. 6.3. Method of solution by the wavelet-based approximation method is presented in Sect. 6.4. In Sect. 6.5, a few numerical experiments are presented. Concluding remarks are given in Sect. 6.6.

6.2 Mathematical Model

A steady-state nonlinear differential equation of substrate concentration based on Michaelis–Menten hypothesis is as follows [2, 8]

$$\frac{\partial^2 c}{\partial x^2} - \frac{lc}{1 + \beta c} = 0, \quad 0 < x \leq 1 \tag{6.1}$$

with the initial conditions

$$c'(0) = 0 \tag{6.2}$$

$$Ac(0) + \beta c(l) + C = 0 \tag{6.3}$$

where c, β, and l denote the dimensionless concentration, saturation parameter, and reaction–diffusion parameter, respectively. Here A, B, and C are constants. Enzyme kinetic problems are highly nonlinear in nature. So, obtaining an exact solution is difficult. However, in this paper more reliable results are obtained with higher computation efficiency.

6.3 Legendre and Chebyshev Wavelets—An Overview

6.3.1 Chebyshev Wavelets

Second kind Chebyshev polynomials are defined in the range $[-1, 1]$ by

$$U_n(x) = \frac{(-2)^n (n+1)!}{(2n+1)!\sqrt{1-x^2}} D^n \left[(1-x^2)^{n+1/2} \right], \tag{6.4}$$

where $D = d/dx$ and it is derived using the recurrence relations (See Ref. [14]).

By varying translation parameter 'b' and dilation parameter 'a' continuously, a family of continuous wavelets can be obtained.

$$\psi_{a,b}(t) = |a|^{-\frac{1}{2}} \psi\left(\frac{t-b}{a}\right) \quad a, b \in R, \ a \neq 0. \tag{6.5}$$

Shifted Second Kind Chebyshev Wavelets They are defined on the interval $[0, 1]$ as

$$\psi_{n,m}(t) = \begin{cases} \frac{2^{\frac{k+3}{2}}}{\sqrt{\pi}} U_m^*(2^k t - n), & t \in \left[\frac{n}{2^k}, \frac{n+1}{2^k}\right] \\ 0 & \text{otherwise} \end{cases} \tag{6.6}$$

$m = 0, 1, \ldots, M$, $n = 0, 1, \ldots, 2^k - 1$. A function $f(t)$ defined over $[0, 1]$ may be expanded in terms second kind Chebyshev wavelets as

$$f(t) = \sum_{n=0}^{\infty} \sum_{m=0}^{\infty} c_{nm} \psi_{nm}(t), \tag{6.7}$$

$$c_{nm} = (f(t), \psi_{nm}(t))_w = \int_0^1 \sqrt{t - t^2} f(t) \psi_{nm}(t) \, dt.$$

If the infinite series is truncated, then it can be written as

$$f(t) = \sum_{n=0}^{\infty} \sum_{m=0}^{\infty} c_{nm} \psi_{nm}(t) = C^T \psi(t), \tag{6.8}$$

where C and $\psi(t)$ are $2^k(M + 1) \times 1$ defined by

$$\left. \begin{array}{l} C = \left[c_{0,0}, c_{0,1}, \ldots c_{0,M}, \ldots, c_{2^k-1,M}, \ldots c_{2^k-1,1}, \ldots, c_{2^k-1,M} \right]^T \\ \psi(t) = \left[\psi_{0,0}, \psi_{0,1}, \ldots, \psi_{0,M}, \ldots \psi_{2^k-1,M}, \ldots, \psi_{2^k-1,1}, \ldots, \psi_{2^k-1,M} \right]^T \end{array} \right\}$$

Theorem 6.3.1 *Let $\Psi(t)$ be the second kind Chebyshev wavelet vector defined in Eq. (6.6). Then the first derivative of the vector $\Psi(t)$ can be expressed as*

$$\frac{d\psi(t)}{dt} = D \psi(t), \tag{6.9}$$

where D is $2^k(M + 1)$ square matrix of derivatives and is defined by

$$D = \begin{bmatrix} F & O & \cdots & O \\ O & F & \cdots & O \\ \vdots & \vdots & \cdots & \vdots \\ O & O & \cdots & F \end{bmatrix}$$

in which F is an $(M + 1)$ square matrix and its (r, s)th element is defined by

$$F_{r,s} = \begin{cases} 2^{k+2}s & r \geq 2, \; r > s \text{ and } (r+s) \text{ odd} \\ 0, & \text{otherwise} \end{cases}. \tag{6.10}$$

Corollary 6.3.1 *The operational matrix for the nth derivative can be obtained from*

$$\frac{d^n \psi(t)}{dt^n} = D^n \psi(t), \quad n = 1, 2, \ldots \text{ where } D^n \text{ is the nth power of } D.$$

6.3.2   Legendre Wavelet Method (LWM)

Legendre wavelets $\psi_{nm}(t) = \psi(k, n, m, t)$ have four arguments: n argument, k can assume any positive integer, m is the order for Legendre polynomials, and t is the normalized time. They are defined on the interval $[0, 1]$ by

$$\psi_{nm}(t) = \begin{cases} \sqrt{\left(m + \frac{1}{2}\right)2^{k+1}} & L_m\left(2^{k+1}t - (2n+1)\right) & \frac{n}{2^k} \le t < \frac{n+1}{2^k} \\ 0 & \text{otherwise} \end{cases} \quad (6.11)$$

where $m = 0, 1, \ldots, M$ and $n = 0, 1, \ldots, 2^k-1$. The coefficient $\sqrt{\left(m + \frac{1}{2}\right)}$ is for orthonormality.

6.4   Method of Solution by the CWM and LWM

First, if we make use of Eqs. (6.1)–(6.3), then the two operational matrices D and D^2 are given by

$$D = \begin{pmatrix} 0 & 0 & 0 \\ 4 & 0 & 0 \\ 0 & 8 & 0 \end{pmatrix} \text{ and } D^2 = \begin{pmatrix} 0 & 0 & 0 \\ 0 & 0 & 0 \\ 32 & 0 & 0 \end{pmatrix}$$

Then, the function $\psi(x)$ can be obtained by using the following relation.

$$\psi(x) = \sqrt{\frac{2}{\pi}} \begin{pmatrix} 2 \\ 8x - 4 \\ 32x^2 - 32x + 6 \end{pmatrix}$$

If we set $C = \left(c_{0,0}, c_{0,1}, c_{0,2}\right)^{T} = \sqrt{\frac{\pi}{2}}(c_0, c_1, c_2)^{T}$, then the shifted second kind Chebyshev wavelet scheme is given by

$$C^{T}D^2\Psi(x) - \frac{1C^{T}\Psi(x)}{1 + \beta C^{T}\Psi(x)} = 0 \quad (6.12)$$

If we select a root $x = \frac{2-\sqrt{2}}{4}$, then we get the following algebraic equation

$$C^{T}D^2\Psi(x) = 64C_2$$
$$C^{T}D\Psi(x) = 8C_1 + 8(8x - 4)C_2$$
$$C^{T}\Psi(x) = 2C_0 + (8x - 4)C_1 + \left(32x^2 - 32x + 6\right)C_2$$

On substituting the boundary condition $C^T D \Psi(0) = 0$, we get

$$64C_2 - \frac{0.1(2C_0 - 2C_2)}{1 + 0.1(2C_0 - 2C_2)} = 0$$

$$64C_2 + 6.092812C_2 + 64C_2^2 - 12.8C_2^2 - 0.095200 - C_2 + 0.2C_2 = 0$$

$$C_0 = 0.482861; \quad C_1 = 0.005488; \quad C_2 = 0.001372$$

$$c = C^T \Psi(x) = 0.043904x^2 + 0.952002 \tag{6.13}$$

Using LWM,

$$D = \begin{bmatrix} 0 & 0 & 0 \\ 2\sqrt{3} & 0 & 0 \\ 0 & 2\sqrt{15} & 0 \end{bmatrix}; \quad D^2 = \begin{bmatrix} 0 & 0 & 0 \\ 0 & 0 & 0 \\ 12\sqrt{5} & 0 & 0 \end{bmatrix};$$

$$\Psi(x) = \sqrt{\frac{2}{\Pi}} \begin{pmatrix} 1 \\ \sqrt{3}(2x - 1) \\ \sqrt{5}(1 - 6x + 6x^2) \end{pmatrix}$$

Then the algebraic expressions are

$$C^T D^2 \Psi(x) = 12\sqrt{5}C_2$$

$$C^T D \Psi(x) = 2\sqrt{3}C_1 + 6\sqrt{5}(2x - 1)C_2$$

$$C^T \Psi(x) = C_0 + \sqrt{3}(2x - 1)C_1 + \sqrt{5}(1 - 6x + 6x^2)C_2$$

On substituting the boundary condition $C^T D \Psi(0) = 0$, we get $C_1 = \sqrt{15}C_2$. Equation (6.1) can be written as

$$C^T D^2 \Psi(x) - \frac{lC^T \Psi(x)}{1 + \beta C^T \Psi(x)} = 0 \tag{6.14}$$

6.5 Numerical Experiments

Limiting Cases:

Case (i): Consider $k = 0.1$ and $\beta = 0.1$. Using the collocation point $x = 1/2$, we obtain

$$12\sqrt{5}C_2 - \frac{0.1\left(2C_0 - \sqrt{5}/2C_2\right)}{1 + 0.1\left(2C_0 - \sqrt{5}/2C_2\right)} = 0$$

$$12\sqrt{5}C_2 + 1.142402\sqrt{5}C_2^2 - 0.0952002 - 0.2\sqrt{5}C_2 + 0.05\sqrt{5}C_2 = 0$$

$$C_0 = 0.966639; \quad C_1 = 0.012676; \quad C_2 = 0.003273$$

$$c = C^{\mathrm{T}}\Psi(x) = 0.043911x^2 + 0.952002 \tag{6.15}$$

Case (ii) Consider Eq. (6.1) with the initial condition [1]

$$C'(0) = z = \left\{ \begin{array}{ll} \mathrm{Sech}\sqrt{l}, & \beta \ll 1 \\ 1 - \frac{l}{2\beta}, & \beta \gg 1 \end{array} \right\} \tag{6.16}$$

The accuracy and efficiency of the proposed wavelet-based methods have been confirmed by using the various values of k. The proposed wavelet-based numerical results have been compared with variational iteration method (VIM) results. For larger k and M, we can get the results very much closer with VIM results [7, 10] (See Figs. 6.1 and 6.2). Error (%) has been calculated using the following formula

$$\mathrm{Error}(\%) = \frac{(\mathrm{Chebyshev\ or\ Legendre}) - \mathrm{VIM}}{\mathrm{VIM}} \times 100 \tag{6.17}$$

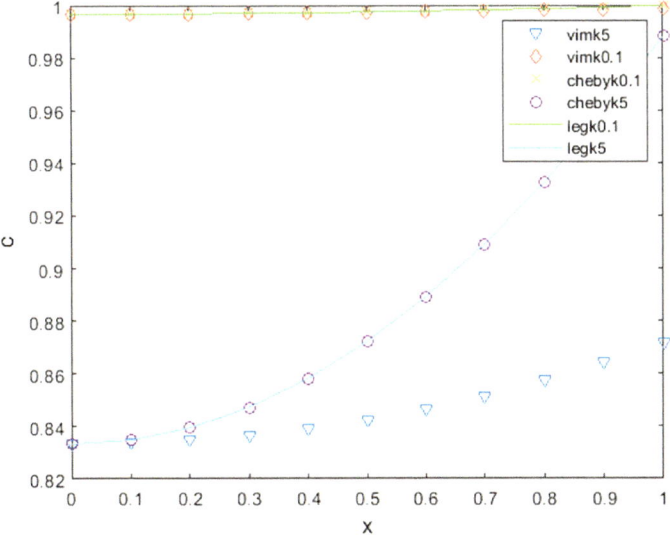

Fig. 6.1 Comparison of CWM, LWM, and VIM for various values of l and β (i) $l = 0.1$, $\beta = 15$ and (ii) $l = 5$, $\beta = 15$

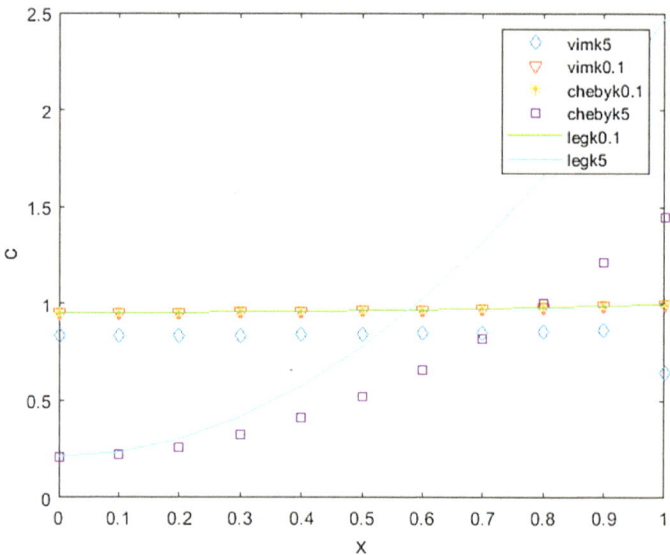

Fig. 6.2 Comparison of CWM, LWM, and VIM for various values of l and β (i) $\beta = 0.1, l = 0.1$ (ii) $\beta = 0.1, l = 5$

Case (iii): Consider Eq. (6.1) with the initial condition [1]

$$\frac{\partial c}{\partial x}\bigg| = 0; \quad c(z) = ls_{\infty} \tag{6.18}$$

The proposed wavelet-based results have been compared with the results obtained by Malvandi and Ganji [7] and Shanmugarajan et al. [13] results. For smaller k, we get the results much closer with the earlier results (See Tables 6.1 and 6.2).

Table 6.1 Numerical solutions of Eq. (6.1) for various values of l and β ($M = 2$)

β	l	CWM	LWM	VIM	% Error CWM	% Error LWM
0.1	0.1	0.087808	0.087822	0.0881	0.331441	0.315550
	5	2.474368	0.452208	2.6343	6.071138	0.828328
15	0.1	0.006208	0.006246	0.0062	0.129032	0.741935
	5	0.309632	0.30965	0.3099	0.086479	0.080671

Table 6.2 Comparison of CWM, LWM, and VIM for various values of l and x ($\beta = 1$)

l	x	CWM	LWM	VIM [7]	% Error CWM	% Error LWM
0.01	0	0.9975	0.9975	0.99750	0.0000	0.0000
	0.2	0.9976	0.9976	0.99760	0.0000	0.0000
	0.4	0.9979	0.9979	0.99790	0.0000	0.0000
	0.6	0.9984	0.9984	0.99840	0.0000	0.0000
	0.8	0.9991	0.9991	0.99910	0.0000	0.0000
	1.0	1.0000	1.0000	1.00000	0.0000	0.0000
1	0	0.7732	0.7732	0.77579	0.33385	0.33385
	0.2	0.7823	0.7823	0.78454	0.28551	0.28551
	0.4	0.8095	0.8095	0.81089	0.17141	0.17141
	0.6	0.8549	0.8549	0.85517	0.03157	0.03157
	0.8	0.9184	0.9184	0.9179	0.00054	0.00054
	1.0	1.0000	1.0000	0.99979	0.02100	0.02100

6.6   Conclusion

Wavelet-based approximation methods have been successfully applied for the steady-state reaction–diffusion problems arising in enzyme kinetics. The obtained numerical results have been compared with VIM and Pade approximation method (PAM) results. The accuracy and simplicity of the proposed wavelet methods have been confirmed by using the various parameter values. The operational matrices contain many zero entries, which lead to the high efficiency of the method and reasonable accuracy is achieved even with less number of collocation points. Another advantage of the proposed wavelet method is that the high accurate numerical solutions are achieved by using small number of the wavelets. It may be concluded that wavelet algorithm is a very powerful and efficient algorithm in finding the approximate solutions for a wide class of linear and nonlinear differential equations.

References

1. K. Indira, L. Rajendran, Analytical expression of the concentration and substrates and product in phenol-polyphenol oxidase system immobilised in laponite hydrogels. Michelis–Menten formalism in homogeneous medium. Electrochim. Acta **56**, 6411–6419 (2011)
2. P.N. Bartlett, R.G. Whitaker, Electrochemical immobilization of enzyme: part-I theory. J. Electroanal. Chem. Interfacial Electrochem. **224**, 27–35 (1987)
3. R. Baronas, F. Ivanauskas, J. Kulys, M. Sapagovas, Modeling of amperometric biosensors with rough surface of the enzyme membrane. J. Math. Chem. **34**, 227–242 (2003)
4. R. Baronas, J. Kulys, F. Ivanauskas, Modelling amperometric enzyme electrode with substrate cyclic conversion. Biosens. Bioelectron. **19**, 915–922 (2004)
5. L. Rajendran, S. Anitha, Reply to "Comments on Analytical solution of amperometric enzymatic reactions based on HPM" by J.-H. He, L.-F. Mo. Electrochim. Acta **102**, 474–476 (2013)
6. J.-H. He, X.-H. Wu, Variational iteration method-new development and applications. Comput. Math. Appl. **54**, 881–894 (2007)

7. A. Malvandi, D.D. Ganji, A general mathematical expression of amperometric enzyme kinetics using He's variational iteration method with Pade approximation. J. Electroanal. Chem. **711**, 32–37 (2013)
8. J.-H. He, L.-F. Mo, Comments on "Analytical solution of amperometric enzymatic reactions based on HPM" by A. Shanmugarajan, S. Alwarappan, S. Somasundaram, R. Lakshmanan. Electrochim. Acta **102**, 472–473 (2013)
9. H. He, Homotopy perturbation technique. Comput. Methods Appl. Mech. Eng. **178**, 257–262 (1999)
10. G. Rahamathunissa, L. Rajendran, Application of He's variational iteration method in nonlinear boundary value problems in enzyme substrate reaction diffusion processes. J. Math. Chem. **44**, 849–861 (2008)
11. A. Meena, L. Rajendran, Mathematical modeling of amperometric and potentiometric biosensors and system of non-linear equations—homotopy perturbation technique. J. Electroanal. Chem. **644**, 50–59 (2010)
12. A. Eswari, L. Rajendran, Analytical expressions of concentration and current in homogeneous catalytic reactions at spherical microelectrodes: homotopy perturbation approach. J. Electroanal. Chem. **651**, 173–184 (2011)
13. A. Shanmugarajan, S. Alwarappan, S. Somasundaram, R. Lakshmanan, Analytical solution of amperometric enzymatic reactions based on homotopy perturbation method. Electrochim. Acta **56**, 3345–3352 (2011)
14. E.H. Doha, A.H. Bhrawy, S.S. Ezz Eldien, Efficient Chebyshev spectral methods for solving multi-term fractional orders differential equations. Appl. Math. Modell. **35**, 5662–5672 (2011)
15. D. Sathyaseelan, G. Hariharan, Wavelet-Based Approximation Algorithms for Some Nonlinear Oscillator Equations Arising in Engineering. J. Inst. Eng. IndiaSer. C (Article in press)

Chapter 7
Haar Wavelet Method for Solving Some Nonlinear Parabolic Equations

Wavelet transform or wavelet analysis is a recently developed mathematical tool in applied mathematics. In this chapter, we develop an accurate and efficient Haar transform or Haar wavelet method for some of the well-known nonlinear parabolic partial differential equations. The equations include the Newell–Whitehead equation, Cahn–Allen equation, FitzHugh–Nagumo equation, Fisher's equation, Burgers' equation, and the Burgers–Fisher equation. The proposed scheme can be used to a wide class of nonlinear equations. The power of this manageable method is confirmed. Moreover, the use of Haar wavelets is found to be accurate, simple, fast, flexible, convenient, small computation costs, and computationally attractive.

7.1 Introduction

Nonlinear phenomena appear in a wide variety of scientific applications such as plasma physics, solid-state physics, optical fibers, biology, fluid dynamics, and chemical kinetics. The concepts like solitons, peakons, kinks, breathers, cusps, and compactons are now thoroughly investigated in the scientific literature [1, 2]. A variety of powerful methods, such as inverse scattering method [3], bilinear transformation [4], Backland transformation, a bilinear form [4], the tanh–sech method [5–7], extended tanh method [6], sine–cosine method [8], homogeneous balance method [9], Exp-function method [10], the tanh method [11], Adomian decomposition method [12], the tanh–coth method [13], Jacobi elliptic functions, and a Lax pair [14] have been used independently by which soliton and multi-soliton solutions are obtained. Recently, Haar wavelets have been applied extensively for signal processing in communications and physics research and have proved to be a wonderful mathematical tool. Haar wavelet method is used to solve some of the nonlinear PDEs given by

© Springer Nature Singapore Pte Ltd. 2019
G. Hariharan, *Wavelet Solutions for Reaction–Diffusion Problems in Science and Engineering*, Forum for Interdisciplinary Mathematics,
https://doi.org/10.1007/978-981-32-9960-3_7

$$u_t = u_{xx} + \alpha u + \beta u^n, \tag{7.1}$$

$$u_t = u_{xx} + \alpha + \beta e^{n\lambda} u, \tag{7.2}$$

$$u_t = u_{xx} - u(1 - u)(\alpha - u), \tag{7.3}$$

$$u_t = u_{xx} + \alpha u u_x, \tag{7.4}$$

$$\text{and} \quad u_t = u_{xx} + \alpha u u_x + ku(1 - u) \tag{7.5}$$

where $\alpha, \beta, k,$ and λ are constants.

In solving ordinary differential equations by using Haar wavelet-related method, Chen and Hsiao [15] had derived an operational matrix of integration based on Haar wavelet. Lepik [16–18] had solved higher-order as well as nonlinear ODEs and some nonlinear evolution equations by Haar wavelet method. Hariharan et al. [19] had introduced the solution of Fisher's equation by Haar wavelet method.

Equation (7.1) gives rise to three known models. For $\alpha = 1, \beta = -1$ and $n = 3$, Eq. (7.1) becomes the Cahn–Allen equation. It arises in many scientific applications such as mathematical biology, quantum mechanics, and plasma physics. It is well known that wave phenomena of plasma media and fluid dynamics are modeled by kink-shaped and tanh solution or bell-shaped sech solutions. The Cahn–Allen equation serves as a model for the study of phase separation in isothermal, isotropic, binary mixtures such as molten alloys. If for $n = 3$ the coefficient β is replaced by $-\beta$, then Eq. (7.1) becomes the Newell–Whitehead equation. The Newell–Whitehead equation describes the dynamical behavior near the bifurcation point for the Rayleigh–Benard convection of binary fluid mixtures [20].

For $n = 2$ and $\beta = -\alpha$, Eq. (7.1) reduces to the well-known Fisher's equation. The Fisher's equation combines diffusion with logistic nonlinearity. This equation is encountered in chemical kinetics and population dynamics, which includes problems such as nonlinear evolution of a population in a one-dimensional habitat, neutron population in a nuclear reaction. Moreover, the same equation occurs in logistic population growth models [21], flame propagation, neurophysiology, autocatalytic chemical reactions, and branching Brownian motion processes. The mathematical properties of Fisher's Equation (FE) have been studied extensively, and there have been numerous discussions in the literature.

Equation (7.2) is a parabolic equation with exponential nonlinearity. Moreover, Eqs. (7.3)–(7.5) give the FitzHugh–Nagumo equation, the Burgers equation, and the Burgers–Fisher equation, respectively. The FitzHugh–Nagumo equation describes the dynamical behavior near the bifurcation point for the Rayleigh–Benard convection of binary fluid mixtures [20]. It is an important nonlinear reaction–diffusion equation and applied to model the transmission of nerve impulses, also used in biology and the area of population genetics, in circuit theory. The Burgers equation, which is a nonlinear partial differential equation of second order, is used in disciplines as a simplified model for turbulence, boundary layer behavior, shock wave formation, and mass transport. The equation serves as a nonlinear analog of the fluid mechanics equations because it has terms, which closely duplicate the physical properties, i.e., a convective term, a diffusive term, and a time-dependent term. Nonlinear partial differential equations (NLPDEs) arise in many fields of science, particularly in physics, engineering, chemistry, and finance and are fundamental for the mathematical formulation of continuum models. The distinctive feature of the Burgers equation is that it is the simplest mathematical formulation of the competition between convection and diffusion. Another feature of the Burgers equation is that although it does not have a pressure gradient term it still is a good approximation of the propagation of one-dimensional disturbances. Performance of a numerical method can be judged from its ability to resolve the large gradient region that develops in the solution. Many problems can be modeled by the Burgers equation. For example, the Burgers equation can be considered as an approach to the Navier–Stokes equation since both contain non-linear terms of the type: unknown functions multiplied by a first derivative and both contain higher-order terms multiplied by a small parameter.

Equations (7.1)–(7.5) arise in many scientific applications such as mathematical biology, quantum mechanics, and plasma physics. Ablowitz and Segur [3] implemented the inverse scattering transform method to handle the nonlinear equations of physical significance where soliton solutions and rational solutions were developed. Mansour [22] showed that traveling wave solutions of a nonlinear reaction–diffusion–chemotaxis model for bacterial pattern formation. Olmos and Shizgal [23] have shown that a pseudo-spectral method of solution of Fisher's equation. Wazwaz [24] showed the analytical study on Burgers, Fisher, and Huxley equations and combined forms of these equations.

We introduce a Haar wavelet method for solving the above equations with the initial and boundary conditions, which will exhibit several advantageous features:

(i) Very high accuracy fast transformation and possibility of implementation of fast algorithms compared with other known methods.
(ii) The simplicity and small computation costs, resulting from the sparsity of the transform matrices and the small number of significant wavelet coefficients.
(iii) The method is also very convenient for solving the boundary value problems, since the boundary conditions are taken care of automatically.

Beginning from 1980s, wavelets have been used for the solution of partial differential equations (PDE). The good features of this approach are possibility to detect singularities, irregular structure, and transient phenomena exhibited by the analyzed equations. Most of the wavelet algorithms can handle exactly periodic boundary conditions. The wavelet algorithms for solving PDE are based on the Galerkin techniques or on the collocation method.

Evidently all attempts to simplify the wavelet solutions for PDE are welcome. One possibility for this is to make use of the Haar wavelet family. Haar wavelets (which are Daubechies of order 1) consist of piecewise constant functions and are therefore the simplest orthonormal wavelets with a compact support. A drawback of the Haar wavelets is their discontinuity. Since the derivatives do not exist in the breaking points, it is not possible to apply the Haar wavelets for solving PDE directly. There are two possibilities for getting out of this situation. One way is to regularize the Haar wavelets with interpolating splines (e.g., B-splines or Deslaurier–Dabuc interpolating wavelets). This approach has been applied by Cattani [25], but the regularization process considerably complicates the solution and the main advantage of the Haar wavelets—the simplicity gets to some extent lost. The other way is to make use of the integral method, which was proposed by Chen and Hsiao [15]. There are discussions by other researchers [26, 27].

7.2　The General Nonlinear Parabolic PDEs

The general nonlinear parabolic equation is of the form

$$u_t = u_{xx} + \alpha u + \beta u^n \tag{7.6}$$

with the initial condition $u(x,0) = f(x), 0 \leq x \leq 1$
and the boundary conditions $u(0,t) = g_0(t), u(1,t) = g_1(t), 0 < t \leq T$.

Let us divide the interval $(0,1]$ into N equal parts of length $\Delta t = (0,1]/N$ and denote $t_s = (s-1)\Delta t$, $s = 1,2,\ldots,N$. We assume that $\dot{u}''(x,t)$ can be expanded in terms of Haar wavelets as formula

$$\dot{u}''(x,t) = \sum_{n=0}^{m-1} c_s(n)h_n(x) = c_{(m)}^T h_{(m)}(x) \tag{7.7}$$

where $\bullet$ and $'$ means differentiation with respect to t and x, respectively, and the row vector $c_{(m)}^T$ is constant in the subinterval $t \in (t_s, t_{s+1}]$.

Integrating Formula (7.7) with respect to t from t_s to t and twice with respect to x from 0 to x, we obtain

$$u''(x,t) = (t - t_s)c_{(m)}^T h_{(m)}(x) + u''(x,t_s) \tag{7.8}$$

$$u(x,t) = (t - t_s)c_{(m)}^T Q_{(m)}h_{(m)}(x) + u(x,t_s) - u(0,t_s) \\ + x[u'(0,t) - u'(0,t_s)] + u(0,t) \tag{7.9}$$

$$\dot{u}(x,t) = c_{(m)}^T Q_{(m)}h_{(m)}(x) + x\dot{u}'(0,t) + \dot{u}(0,t) \tag{7.10}$$

By the boundary conditions, we obtain

$$u(0, t_s) = g_0(t_s), \quad u(1, t_s) = g_1(t_s)$$
$$\dot{u}(0, t) = g_0'(t), \quad \dot{u}(1, t) = g_1'(t)$$

Putting $x = 1$ in Formulae (7.9) and (7.10), we have

$$u'(0, t) - u'(0, t_s) = -(t - t_s) c_{(m)}^T Q_{(m)} h_{(m)}(x)$$
$$+ g_1(t) - g_0(t) - g_1(t_s) + g_0(t_s) \tag{7.11}$$

$$\dot{u}'(0, t) = g_1'(t) - c_{(m)}^T Q_{(m)} h_{(m)}(x) - g_0'(t) \tag{7.12}$$

Substituting Formulae (7.11) and (7.12) into Formulae (7.8)–(7.10) and discretizing the results by assuming $x \to x_l$, $t \to t_{s+1}$, we obtain

$$u''(x_l, t_{s+1}) = (t_{s+1} - t_s) c_{(m)}^T h_{(m)}(x_l) + u''(x_l, t_s) \tag{7.13}$$

$$u(x_l, t_{s+1}) = (t_{s+1} - t_s) c_{(m)}^T Q_{(m)} h_{(m)}(x_l) + u(x_l, t_s) - g_0(t_s) + g_0(t_{s+1})$$
$$+ x_l \left[-(t_{s+1} - t_s) c_{(m)}^T P_{(m)} f + g_1(t_{s+1}) - g_0(t_{s+1}) - g_1(t_s) + g_0(t_s) \right] \tag{7.14}$$

$$\dot{u}(x_l, t_{s+1}) = c_{(m)}^T Q_{(m)} h_{(m)}(x) + g_0'(t_{s+1}) + x_l \left[-c_{(m)}^T P_{(m)} f + g_1'(t_{s+1}) - g_0'(t_{s+1}) \right], \tag{7.15}$$

where the vector f is defined as

$$f = [1, \underbrace{0, \ldots, 0}_{(m-1) \text{ elements}}]^T$$

In the following, the scheme

$$\dot{u}(x_l, t_{s+1}) = u''(x_l, t_{s+1}) + \alpha u(x_l, t_{s+1}) + \beta u''(x_l, t_{s+1}) \tag{7.16}$$

which leads us from the time layer t_s to t_{s+1} is used.
Substituting Eqs. (7.13)–(7.15) into the Eq. (7.16), we gain

$$c_{(m)}^T Q_{(m)} h_{(m)}(x_l) + x_l \left[-c_{(m)}^T P_{(m)} f + g_1'(t_{s+1}) - g_0'(t_{s+1}) \right] + g_0'(t_{s+1})$$
$$= u''(x_l, t_{s+1}) + \alpha u(x_l, t_{s+1}) + \beta u''(x_l, t_{s+1}) \tag{7.17}$$

From Formula (7.17), the wavelet coefficients $c_{(m)}^T$ can be successively calculated.

7.3   Parabolic Equation with Exponential Nonlinearity

As stated before, this equation is defined by

$$u_t = u_{xx} + \alpha + \beta e^{n\lambda u} \tag{7.18}$$

with the initial condition $u(x,0) = f(x), \ 0 \le x \le 1$
and the boundary conditions $u(0,t) = g_0(t), \ u(1,t) = g_1(t), \ 0 < t \le T$.

Let us divide the interval $(0,1]$ into N equal parts of length $\Delta t = (0,1]/N$ and denote $t_s = (s-1)\Delta t, \ s = 1, 2, \ldots, N$. We assume that $\dot{u}''(x,t)$ can be expanded in terms of Haar wavelets as formula

$$\dot{u}''(x,t) = \sum_{n=0}^{m-1} c_s(n) h_n(x) = c_{(m)}^{\mathrm{T}} h_{(m)}(x) \tag{7.19}$$

where $\cdot$ and $'$ means differentiation with respect to t and x, respectively, and the row vector $c_{(m)}^{\mathrm{T}}$ is constant in the subinterval $t \in (t_s, t_{s+1}]$.

Integrating Formula (7.19) with respect to t from t_s to t and twice with respect to x from 0 to x, we obtain

$$u''(x,t) = (t - t_s) c_{(m)}^{\mathrm{T}} h_{(m)}(x) + u''(x,t_s) \tag{7.20}$$

$$\begin{aligned} u(x,t) = {}& (t - t_s) c_{(m)}^{\mathrm{T}} Q_{(m)} h_{(m)}(x) + u(x,t_s) - u(0,t_s) \\ & + x[u'(0,t) - u'(0,t_s)] + u(0,t) \end{aligned} \tag{7.21}$$

$$\dot{u}(x,t) = c_{(m)}^{\mathrm{T}} Q_{(m)} h_{(m)}(x) + x\dot{u}'(0,t) + \dot{u}(0,t) \tag{7.22}$$

By the boundary conditions, we obtain

$$\begin{aligned} u(0,t_s) &= g_0(t_s), \quad u(1,t_s) = g_1(t_s) \\ \dot{u}(0,t) &= g_0'(t), \quad \dot{u}(1,t) = g_1'(t) \end{aligned}$$

Putting $x = 1$ in Formulae (7.21) and (7.22), we have

$$\begin{aligned} u'(0,t) - u'(0,t_s) = {}& -(t - t_s) c_{(m)}^{\mathrm{T}} Q_{(m)} h_{(m)}(x) \\ & + g_1(t) - g_0(t) - g_1(t_s) + g_0(t_s) \end{aligned} \tag{7.23}$$

$$\dot{u}'(0,t) = g_1'(t) - c_{(m)}^{\mathrm{T}} Q_{(m)} h_{(m)}(x) - g_0'(t) \tag{7.24}$$

Substituting Formulae (7.23) and (7.24) into Formulae (7.20)–(7.22) and discretizing the results by assuming $x \to x_l, \ t \to t_{s+1}$, we obtain

$$u''(x_l, t_{s+1}) = (t_{s+1} - t_s)c_{(m)}^{\mathrm{T}} h_{(m)}(x_l) + u''(x_l, t_s) \tag{7.25}$$

$$u(x_l, t_{s+1}) = (t_{s+1} - t_s)c_{(m)}^{\mathrm{T}} Q_{(m)} h_{(m)}(x_l) + u(x_l, t_s) - g_0(t_s) + g_0(t_{s+1})$$
$$+ x_l \left[-(t_{s+1} - t_s)c_{(m)}^{\mathrm{T}} P_{(m)} f + g_1(t_{s+1}) - g_0(t_{s+1}) - g_1(t_s) + g_0(t_s) \right] \tag{7.26}$$

$$\dot{u}(x_l, t_{s+1}) = c_{(m)}^{\mathrm{T}} Q_{(m)} h_{(m)}(x) + g_0'(t_{s+1}) + x_l \left[-c_{(m)}^{\mathrm{T}} P_{(m)} f + g_1'(t_{s+1}) - g_0'(t_{s+1}) \right] \tag{7.27}$$

where the vector f is defined as

$$f = [1, \underbrace{0, \ldots, 0}_{(m-1)\ \text{elements}}]^{\mathrm{T}}$$

In the following, the scheme

$$\dot{u}(x_l, t_{s+1}) = u''(x_l, t_{s+1}) + \alpha + \beta e^{n\lambda u(x_l, t_{s+1})} \tag{7.28}$$

which leads us from the time layer t_s to t_{s+1} is used.

Substituting Eqs. (7.25)–(7.27) into the Eq. (7.28), we gain

$$c_{(m)}^{\mathrm{T}} Q_{(m)} h_{(m)}(x_l) + x_l \left[-c_{(m)}^{\mathrm{T}} P_{(m)} f + g_1'(t_{s+1}) - g_0'(t_{s+1}) \right] + g_0'(t_{s+1})$$
$$= u''(x_l, t_{s+1}) + \alpha + \beta e^{n\lambda u(x_l, t_{s+1})} \tag{7.29}$$

From Formula (7.29), the wavelet coefficients $c_{(m)}^{\mathrm{T}}$ can be successively calculated.

7.4 The FitzHugh–Nagumo Equation

We consider the FitzHugh–Nagumo equation

$$u_t = u_{xx} - u(1 - u)(\alpha - u) \tag{7.30}$$

with the initial condition $u(x, 0) = f(x)$, $0 \leq x \leq 1$
and the boundary conditions $u(0, t) = g_0(t)$, $u(1, t) = g_1(t)$, $0 < t \leq T$.

Let us divide the interval $(0, 1]$ into N equal parts of length $\Delta t = (0, 1]/N$ and denote $t_s = (s - 1)\Delta t$, $s = 1, 2, \ldots, N$. We assume that $\dot{u}''(x, t)$ can be expanded in terms of Haar wavelets as formula

$$\ddot{u}''(x,t) = \sum_{n=0}^{m-1} c_s(n)h_n(x) = c_{(m)}^{\mathrm{T}} h_{(m)}(x) \tag{7.31}$$

where $\cdot$ and $'$ means differentiation with respect to t and x, respectively, and the row vector $c_{(m)}^{\mathrm{T}}$ is constant in the subinterval $t \in (t_s, t_{s+1}]$.

Integrating Formula (7.31) with respect to t from t_s to t and twice with respect to x from 0 to x, we obtain

$$u''(x,t) = (t - t_s)c_{(m)}^{\mathrm{T}} h_{(m)}(x) + u''(x,t_s) \tag{7.32}$$

$$\begin{aligned} u(x,t) = (t - t_s)c_{(m)}^{\mathrm{T}} Q_{(m)}h_{(m)}(x) + u(x,t_s) - u(0,t_s) \\ + x[u'(0,t) - u'(0,t_s)] + u(0,t) \end{aligned} \tag{7.33}$$

$$\dot{u}(x,t) = c_{(m)}^{\mathrm{T}} Q_{(m)}h_{(m)}(x) + x\dot{u}'(0,t) + \dot{u}(0,t) \tag{7.34}$$

By the boundary conditions, we obtain

$$\begin{aligned} u(0,t_s) &= g_0(t_s), & u(1,t_s) &= g_1(t_s) \\ \dot{u}(0,t) &= g_0'(t), & \dot{u}(1,t) &= g_1'(t) \end{aligned}$$

Putting $x = 1$ in Formulae (7.33) and (7.34), we have

$$\begin{aligned} u'(0,t) - u'(0,t_s) = -(t - t_s)c_{(m)}^{\mathrm{T}} Q_{(m)}h_{(m)}(x) \\ + g_1(t) - g_0(t) - g_1(t_s) + g_0(t_s) \end{aligned} \tag{7.35}$$

$$\dot{u}'(0,t) = g_1'(t) - c_{(m)}^{\mathrm{T}} Q_{(m)}h_{(m)}(x) - g_0'(t) \tag{7.36}$$

Substituting Formulae (7.35) and (7.36) into Formulae (7.32)–(7.34) and discretizing the results by assuming $x \to x_l$, $t \to t_{s+1}$, we obtain

$$u''(x_l, t_{s+1}) = (t_{s+1} - t_s)c_{(m)}^{\mathrm{T}} h_{(m)}(x_l) + u''(x_l, t_s) \tag{7.37}$$

$$\begin{aligned} u(x_l, t_{s+1}) = (t_{s+1} - t_s)c_{(m)}^{\mathrm{T}} Q_{(m)}h_{(m)}(x_l) + u(x_l, t_s) - g_0(t_s) + g_0(t_{s+1}) \\ + x_l \Big[-(t_{s+1} - t_s)c_{(m)}^{\mathrm{T}} P_{(m)}f + g_1(t_{s+1}) - g_0(t_{s+1}) - g_1(t_s) + g_0(t_s) \Big] \end{aligned} \tag{7.38}$$

$$\dot{u}(x_l, t_{s+1}) = c_{(m)}^{\mathrm{T}} Q_{(m)}h_{(m)}(x) + g_0'(t_{s+1}) + x_l \Big[-c_{(m)}^{\mathrm{T}} P_{(m)}f + g_1'(t_{s+1}) - g_0'(t_{s+1}) \Big] \tag{7.39}$$

where the vector f is defined as

$$f = [1, \underbrace{0, \ldots, 0}_{(m-1)\ \text{elements}}]^{\mathrm{T}}$$

In the following, the scheme

$$\dot{u}(x_l, t_{s+1}) = u''(x_l, t_{s+1}) - u(x_l, t_{s+1})[1 - u(x_l, t_{s+1})][\alpha - u(x_l, t_{s+1})] \quad (7.40)$$

which leads us from the time layer t_s to t_{s+1} is used.

Substituting Eqs. (7.37)–(7.39) into the Eq. (7.40), we gain

$$\begin{aligned}
& c_{(m)}^{\mathrm{T}} Q_{(m)} h_{(m)}(x_l) + x_l \left[-c_{(m)}^{\mathrm{T}} P_{(m)} f + g_1'(t_{s+1}) - g_0'(t_{s+1}) \right] + g_0'(t_{s+1}) \\
& = u''(x_l, t_{s+1}) - u(x_l, t_{s+1})[1 - u(x_l, t_{s+1})][\alpha - u(x_l, t_{s+1})]
\end{aligned} \quad (7.41)$$

From Formula (7.41), the wavelet coefficients $c_{(m)}^{\mathrm{T}}$ can be successively calculated.

7.5 The Burgers Equation

We consider the Burgers equation

$$u_t = u_{xx} + \alpha u u_x \quad (7.42)$$

with the initial condition $u(x, 0) = f(x), \ 0 \le x \le 1$
and the boundary conditions $u(0, t) = g_0(t), \ u(1, t) = g_1(t), \ 0 < t \le T$.

Let us divide the interval $(0, 1]$ into N equal parts of length $\Delta t = (0, 1]/N$ and denote $t_s = (s - 1)\Delta t, \ s = 1, 2, \ldots, N$. We assume that $\dot{u}''(x, t)$ can be expanded in terms of Haar wavelets as formula

$$\dot{u}''(x, t) = \sum_{n=0}^{m-1} c_s(n) h_n(x) = c_{(m)}^{\mathrm{T}} h_{(m)}(x) \quad (7.43)$$

where $\cdot$ and $'$ means differentiation with respect to t and x, respectively, and the row vector $c_{(m)}^{\mathrm{T}}$ is constant in the subinterval $t \in (t_s, t_{s+1}]$.

Integrating Formula (7.43) with respect to t from t_s to t and twice with respect to x from 0 to x, we obtain

$$u''(x,t) = (t - t_s)c_{(m)}^T h_{(m)}(x) + u''(x,t_s) \tag{7.44}$$

$$u'(x,t) = (t - t_s)c_{(m)}^T P_{(m)} h_{(m)}(x) + u'(x,t_s) - u'(0,t_s) + u'(0,t) \tag{7.45}$$

$$\begin{aligned} u(x,t) = (t - t_s)c_{(m)}^T Q_{(m)} h_{(m)}(x) + u(x,t_s) - u(0,t_s) \\ + x[u'(0,t) - u'(0,t_s)] + u(0,t) \end{aligned} \tag{7.46}$$

$$\dot{u}(x,t) = c_{(m)}^T Q_{(m)} h_{(m)}(x) + x\dot{u}'(0,t) + \dot{u}(0,t) \tag{7.47}$$

By the boundary conditions, we obtain

$$\begin{aligned} u(0,t_s) = g_0(t_s), \quad u(1,t_s) = g_1(t_s) \\ \dot{u}(0,t) = g_0'(t), \quad \dot{u}(1,t) = g_1'(t) \end{aligned}$$

Putting $x = 1$ in Formulae (7.46) and (7.47), we have

$$\begin{aligned} u'(0,t) - u'(0,t_s) = -(t - t_s)c_{(m)}^T Q_{(m)} h_{(m)}(x) \\ + g_1(t) - g_0(t) - g_1(t_s) + g_0(t_s) \end{aligned} \tag{7.48}$$

$$\dot{u}'(0,t) = g_1'(t) - c_{(m)}^T Q_{(m)} h_{(m)}(x) - g_0'(t) \tag{7.49}$$

Substituting Formulae (7.48) and (7.49) into Formulae (7.44)–(7.47) and discretizing the results by assuming $x \to x_l$, $t \to t_{s+1}$, we obtain

$$u''(x_l, t_{s+1}) = (t_{s+1} - t_s)c_{(m)}^T h_{(m)}(x_l) + u''(x_l, t_s) \tag{7.50}$$

$$u'(x_l, t_{s+1}) = (t_{s+1} - t_s)c_{(m)}^T P_{(m)} h_{(m)}(x_l) + u'(x_l, t_s) - (t_{s+1} - t_s)c_{(m)}^T P_{(m)} f \tag{7.51}$$

$$\begin{aligned} u(x_l, t_{s+1}) = (t_{s+1} - t_s)c_{(m)}^T Q_{(m)} h_{(m)}(x_l) + u(x_l, t_s) - g_0(t_s) + g_0(t_{s+1}) \\ + x_l \left[-(t_{s+1} - t_s)c_{(m)}^T P_{(m)} f + g_1(t_{s+1}) - g_0(t_{s+1}) - g_1(t_s) + g_0(t_s) \right] \end{aligned} \tag{7.52}$$

$$\dot{u}(x_l, t_{s+1}) = c_{(m)}^T Q_{(m)} h_{(m)}(x) + g_0'(t_{s+1}) + x_l \left[-c_{(m)}^T P_{(m)} f + g_1'(t_{s+1}) - g_0'(t_{s+1}) \right], \tag{7.53}$$

where the vector f is defined as

$$f = [1, \quad \underbrace{0, \ldots, 0}_{(m-1) \text{ elements}}]^T$$

In the following, the scheme

$$\dot{u}(x_l, t_{s+1}) = u''(x_l, t_{s+1}) + \alpha u(x_l, t_{s+1}) u'(x_l, t_{s+1}) \tag{7.54}$$

which leads us from the time layer t_s to t_{s+1} is used.

Substituting Eqs. (7.50)–(7.53) into the Eq. (7.54), we gain

$$c_{(m)}^T Q_{(m)} h_{(m)}(x_l) + x_l \left[-c_{(m)}^T P_{(m)} f + g_1'(t_{s+1}) - g_0'(t_{s+1}) \right] + g_0'(t_{s+1})$$
$$= u''(x_l, t_{s+1}) + \alpha u(x_l, t_{s+1}) u'(x_l, t_{s+1}) \tag{7.55}$$

From Formula (7.55), the wavelet coefficients $c_{(m)}^T$ can be successively calculated.

7.6 The Burgers–Fisher Equation

We consider the Burgers–Fisher equation

$$u_t = u_{xx} + \alpha u u_x + ku(1-u) \tag{7.56}$$

with the initial condition $u(x,0) = f(x), \quad 0 \le x \le 1$
and the boundary conditions $u(0,t) = g_0(t), \, u(1,t) = g_1(t), \quad 0 < t \le T$.

Let us divide the interval $(0,1]$ into N equal parts of length $\Delta t = (0,1]/N$ and denote $t_s = (s-1)\Delta t, \, s = 1, 2, \ldots, N$. We assume that $\dot{u}''(x,t)$ can be expanded in terms of Haar wavelets as formula

$$\dot{u}''(x,t) = \sum_{n=0}^{m-1} c_s(n) h_n(x) = c_{(m)}^T h_{(m)}(x) \tag{7.57}$$

where $\cdot$ and $'$ means differentiation with respect to t and x, respectively, and the row vector $c_{(m)}^T$ is constant in the subinterval $t \in (t_s, t_{s+1}]$.

Integrating Formula (7.57) with respect to t from t_s to t and twice with respect to x from 0 to x, we obtain

$$u''(x,t) = (t - t_s)c_{(m)}^T h_{(m)}(x) + u''(x, t_s) \tag{7.58}$$

$$u'(x,t) = (t - t_s)c_{(m)}^T P_{(m)} h_{(m)}(x) + u'(x, t_s) - u'(0, t_s) + u'(0, t) \tag{7.59}$$

$$u(x,t) = (t - t_s)c_{(m)}^T Q_{(m)} h_{(m)}(x) + u(x, t_s) - u(0, t_s)$$
$$+ x[u'(0,t) - u'(0,t_s)] + u(0,t) \tag{7.60}$$

$$\dot{u}(x,t) = c_{(m)}^T Q_{(m)} h_{(m)}(x) + x\dot{u}'(0,t) + \dot{u}(0,t) \tag{7.61}$$

By the boundary conditions, we obtain

$$u(0,t_s) = g_0(t_s), \quad u(1,t_s) = g_1(t_s)$$
$$\dot{u}(0,t) = g'_0(t), \quad \dot{u}(1,t) = g'_1(t)$$

Putting $x = 1$ in Formulae (7.60) and (7.61), we have

$$u'(0,t) - u'(0,t_s) = -(t - t_s)c_{(m)}^T Q_{(m)} h_{(m)}(x)$$
$$+ g_1(t) - g_0(t) - g_1(t_s) + g_0(t_s) \tag{7.62}$$

$$\dot{u}'(0,t) = g'_0(t) - c_{(m)}^T Q_{(m)} h_{(m)}(x) - g'_0(t) \tag{7.63}$$

Substituting Formulae (7.62) and (7.64) into Formulae (7.58)–(7.61) and discretizing the results by assuming $x \to x_l$, $t \to t_{s+1}$, we obtain

$$u''(x_l, t_{s+1}) = (t_{s+1} - t_s)c_{(m)}^T h_{(m)}(x_l) + u''(x_l, t_s) \tag{7.64}$$

$$u'(x_l, t_{s+1}) = (t_{s+1} - t_s)c_{(m)}^T P_{(m)} h_{(m)}(x_l) + u'(x_l, t_s) - (t_{s+1} - t_s)c_{(m)}^T P_{(m)} f \tag{7.65}$$

$$u(x_l, t_{s+1}) = (t_{s+1} - t_s)c_{(m)}^T Q_{(m)} h_{(m)}(x_l) + u(x_l, t_s) - g_0(t_s) + g_0(t_{s+1})$$
$$+ x_l \left[-(t_{s+1} - t_s)c_{(m)}^T P_{(m)} f + g_1(t_{s+1}) - g_0(t_{s+1}) - g_1(t_s) + g_0(t_s) \right] \tag{7.66}$$

$$\dot{u}(x_l, t_{s+1}) = c_{(m)}^T Q_{(m)} h_{(m)}(x) + g'_0(t_{s+1}) + x_l \left[-c_{(m)}^T P_{(m)} f + g'_1(t_{s+1}) - g'_0(t_{s+1}) \right] \tag{7.67}$$

where the vector f is defined as

$$f = [1, \quad \underbrace{0, \ldots, 0}_{(m-1) \text{ elements}}]^{\mathrm{T}}$$

In the following, the scheme

$$\dot{u}(x_l, t_{s+1}) = u''(x_l, t_{s+1}) + \alpha u(x_l, t_{s+1}) u'(x_l, t_{s+1}) + ku(x_l, t_{s+1})[1 - u(x_l, t_{s+1})] \tag{7.68}$$

which leads us from the time layer t_s to t_{s+1} is used.

Substituting Eqs. (7.50)–(7.53) into the Eq. (7.54), we gain

$$c_{(m)}^{\mathrm{T}} Q_{(m)} h_{(m)}(x_l) + x_l \left[-c_{(m)}^{\mathrm{T}} P_{(m)} f + g_1'(t_{s+1}) - g_0'(t_{s+1}) \right] + g_0'(t_{s+1}) \tag{7.69}$$
$$= u''(x_l, t_{s+1}) + \alpha u(x_l, t_{s+1}) u'(x_l, t_{s+1}) + ku(x_l, t_{s+1})[1 - u(x_l, t_{s+1})]$$

From Formula (7.69), the wavelet coefficients $c_{(m)}^{\mathrm{T}}$ can be successively calculated.

Computer simulation was carried out in the cases $m = 32$, $m = 64$, $m = 128$, and $m = 256$; the computed results were compared with the exact solution; more accurate results can be obtained by using a larger m. The fast capability of the Haar transform is impressive. Figure 7.1 shows the comparison between the Haar and exact solutions for $m = 16$. For smaller m, the results coincide with the exact solution.

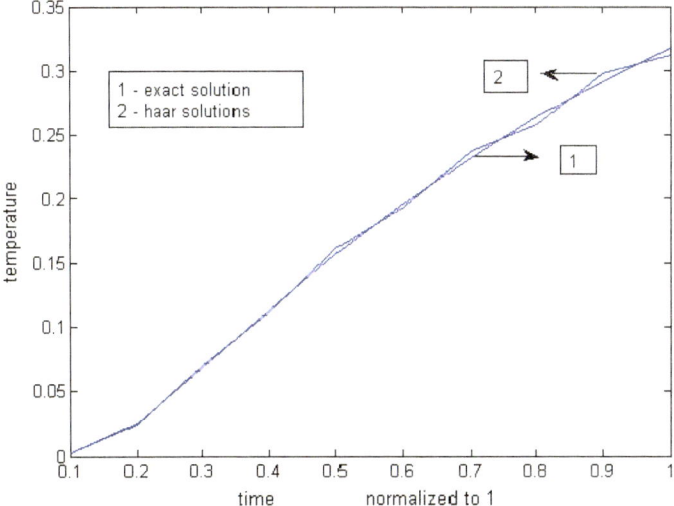

Fig. 7.1 Comparison between exact and Haar solution of Fisher's equation $x = 20$ and $k = 12.5$

7.7   Conclusion

The theoretical elegance of the Haar wavelet approach can be appreciated from the simple mathematical relations and their compact derivations and proofs. It has been well demonstrated that in applying the nice properties of Haar wavelets, the differential equations can be solved conveniently and accurately by using the Haar wavelet method systematically. In comparison with the existing numerical schemes used to solve the nonlinear parabolic equations, the scheme in this paper is an improvement over other methods in terms of accuracy. It is worth mentioning that Haar solution provides excellent results even for small values of m ($m = 16$). For larger values of m (i.e., $m = 32$, $m = 64$, $m = 128$, $m = 256$), we can obtain the results closer to the real values. The method with far less degrees of freedom and with smaller CPU time provides better solutions than classical ones. The main goal of this work is to apply the Haar wavelet method to some well-known nonlinear parabolic equations that appear in many scientific applications. The work also confirmed the power of the Haar wavelet method in handling nonlinear equations in general. This method can be easily extended to find the solution of all other nonlinear parabolic equations. Another benefit of our method is that the scheme presented here, with some modifications, seems to be easily extended to solve model equations including more mechanical, physical, or biophysical effects, such as nonlinear convection, reaction, linear diffusion, and dispersion.

References

1. M. Wadati, Introduction to solitons. Pramana J. Phys. **57**(5–6), 841–847 (2001)
2. M. Wadati, The modified Korteweg-de Veries equation. J. Phys. Soc. Jpn. **34**, 1289–1296 (1973)
3. M. Ablowitz, H. Segur, *Solitons and the Inverse Scattering Transform* (SIAM, Philadelphia, 1981)
4. R. Hirota, *Direct Methods in Soliton Theory* (Springer, Berlin, 1980)
5. W. Malfliet, W. Hereman, The tanh method I: exact solutions of nonlinear evolution and wave equations. Phys. Scr. **54**, 563 (1996)
6. A.M. Wazwaz, The extended tanh method for abundant solitary wave solutions of nonlinear wave equations. Appl. Math. Comput. **187**, 1131 (2007)
7. E. Yusufoğlu, A. Bekir, Exact solutions of coupled nonlinear evolution equations. Chaos, Solitons Fractals **37**(3), 842 (2008)
8. A.M. Wazwaz, A sine–cosine method for handling nonlinear wave equations. Math. Comput. Model. **40**, 499 (2004)
9. E. Fan, H. Zhang, A note on the homogeneous balance method. Phys. Lett. A **246**, 403 (1998)
10. S. Zhang, Application of exp-function method to a KdV equation with variable coefficients. Phys. Lett. A **365**, 448 (2007)
11. A.M. Wazwaz, The tanh method for travelling wave solutions of nonlinear equations. Appl. Math. Comput. **154**(3), 713 (2004)
12. A.M. Wazwaz, An analytical study of Fisher's equation by using Adomian decomposition method. Appl. Math. Comput. **154**, 609–620 (2004)

13. A.M. Wazwaz, The tanh–coth method for solitons and kink solutions for nonlinear parabolic equations. J. Appl. Math. Comput. **188**, 1467 (2007)
14. P.D. Lax, Integrals of nonlinear equations of evolution and solitary waves. Comm. Pure Appl. Math. **62**, 467–490 (1968)
15. C.F. Chen, C.H. Hsiao, Haar wavelet method for solving lumped and distributed-parameter systems. IEEE Proc. Pt. D **144**(1), 87–94 (1997)
16. U. Lepik, Numerical solution of evolution equations by the Haar wavelet method. J. Appl. Math. Comput. **185**, 695–704 (2007)
17. U. Lepik, Numerical solution of differential equations using Haar wavelets. Math. Comp. Simul. **68**, 127–143 (2005)
18. U. Lepik, Application of the Haar wavelet transform to solving integral and differential equations. Proc. Est. Acad. Sci. Phys. Math. **56**(1), 28–46 (2007)
19. G. Hariharan, K. Kannan, K.R. Sharma, Haar wavelet method for solving Fisher's equation. Appl. Math. Comput. **211**(2), 284–292 (2009)
20. H.C. Rosu, O. Cornejo-Pe'rez, Super symmetric pairing of kinks for polynomial nonlinearities. Phys. Rev. E **71**, 1–13 (2005)
21. P. Brazhnik, J. Tyson, On traveling wave solutions of Fisher's equation in two spatial dimensions. SIAM J. Appl. Math. **60**(2), 371–391 (1999)
22. M.B.A. Monsour, Travelling wave solutions of a nonlinear reaction-diffusion-chemotaxis model for bacterial pattern formation. Appl. Math. Model. **32**, 240–247 (2008)
23. D. Olmos, B.D. Shizgal, A pseudospectral method of solution of Fisher's equation. J. Comput. Appl. Math. **193**, 219–242 (2006)
24. A.M. Wazwaz, Analytical study on Burgers, Fisher, Huxley equations and combined forms of these equations. J. Appl. Math. Comput. **195**, 754–761 (2008)
25. C. Cattani, Haar wavelet spline. J. Interdisc. Math. **4**, 35–47 (2001)
26. G. Hariharan, K. Kannan, K.R. Sharma, Haar wavelet in estimating depth profile of soil temperature. Appl. Math. Comput. **210**(1), 119–125 (2009)
27. C.H. Hsiao, W.J. Wang, Haar wavelet approach to nonlinear stiff systems. Math. Comput. Simul. **57**, 347–353 (2001)

Chapter 8
An Efficient Wavelet-Based Approximation Method to Gene Propagation Model Arising in Population Biology

In this chapter, we have applied an efficient wavelet-based approximation method for solving Fisher's type and fractional Fisher's type equations arising in biological sciences. To the best of our knowledge, until now there is no rigorous wavelet solution has been addressed for Fisher's and fractional Fisher's equations. The highest derivative in the differential equation is expanded into Legendre series, and this approximation is integrated while the boundary conditions are applied using integration constants. With the help of Legendre wavelets operational matrices, Fisher's equation and fractional Fisher's equation are converted into a system of algebraic equations. Block-pulse functions are used to investigate the Legendre wavelets coefficient vectors of nonlinear terms. The convergence of the proposed methods is proved. Finally, we have given some numerical examples to demonstrate the validity and applicability of the method.

8.1 Introduction

Wavelet analysis, as a relatively new area in applied mathematical research, has received considerable attention in dealing with PDEs and fractional-type PDEs [1–3]. The propagation of a mutant gene model was first introduced by Fisher, which is known as Fisher's equation [4]. These equations have wide applications in the fields of logistic population growth, flame propagation, euro physiology, autocatalytic chemical reactions, branching Brownian motion processes, and nuclear reactor theory [2, 5, 6]. The Fisher–Kolmogorov equation describes the growth of a gene within a population. We have seen that the solution can easily be described as a traveling wave-moving with constant speed and without change of the front's shape. This means that the growth of the gene is the same at very time. We have used the leading edge approximation to the asymptotic behavior of left- and right-moving fronts. With an asymmetric derivative, we obtain different properties for both directions of propagation. The right-moving front is accelerated,

© Springer Nature Singapore Pte Ltd. 2019
G. Hariharan, *Wavelet Solutions for Reaction–Diffusion Problems in Science and Engineering*, Forum for Interdisciplinary Mathematics,
https://doi.org/10.1007/978-981-32-9960-3_8

and again, the leading edge approximation has permitted to calculate the exponential speed. As for the Fisher–Kolmogorov model, we observe that for decreasing values of α, profiles accelerate later and at the beginning of the simulation, profiles seem to move with constant speed.

In recent years, wavelet transforms have found their way into many different fields in science and engineering. Moreover, wavelets established a connection with fast numerical algorithms.

Wavelet theory possesses many useful properties, such as compact support, orthogonality, dyadic, orthonormality, and multi-resolution analysis (MRA). Fractional partial differential equations (FPDEs) are generalizations of classical partial differential equations of integer order. Mathematical modeling of complex processes is a major challenge for contemporary scientist. In contrast to simple classical systems, where the theory of integer order differential equations is sufficient to describe their dynamics, fractional derivatives provide an excellent and an efficient instrument for the description of memory and hereditary properties of various complex materials and systems [7–13]. But these FPDEs are difficult to get their exact solutions [7–9, 14]. So the approximation methods must be used. Analytical methods enable researchers to study the effect of different variables or parameters on the function under study easily. Recently, there are several new approaches have been proposed for solving nonlinear PDEs, for example, the Adomian decomposition method [5, 15], the variational iteration method [16], differential transform method [17], reduced differential transform method [18], homotopy analysis method [19–21], and exp-function method [22]. Recently, local fractional calculus has been used to deal with problems for nondifferentiable functions; see [23–26] and the references therein. Local fractional Fourier series method is one of very efficient and powerful techniques for finding the solutions of the local fractional differential equations. It is also worth noting that the advantage of the local fractional differential equations displays the nondifferential solutions, which show the fractal and local behaviors of moments.

In recent years, nonlinear reaction–diffusion equations (NLRDE) have been used as a basis for a wide variety of models, for the special spread of gene in population [13] and for chemical wave propagation [2, 3]. Wazwaz and Gorguis [5] developed the Adomian decomposition method for the Fisher-type equations. Carey and Shen [27] implemented the least-squares finite element method for Fisher's reaction–diffusion equation. Al-khaled [28] introduced the sinc-collocation method by the pseudo-spectral method for the numerical solution of Fisher's equation. Mittal and Jiwari [29] have presented the differential quadrature method for Fisher's equations. Merdan [30] solved the time-fractional reaction–diffusion equations by the fractional variational iteration method. Khan et al. [19] established the analytical solutions of the fractional reaction–diffusion equations by the homotopy analysis method. Kurulay and Bayram [31] showed the numerical solutions of time-fractional reaction–diffusion equation by the differential transform method. Al-Min Yang et al. [32] addressed a transient heat conduction problem in a fractal semi-infinite bar solved by the Yang–Fourier transform.

In the numerical analysis, wavelet-based methods and hybrid methods become important tools because of the properties of localization. In wavelet-based methods, there are two important ways of improving the approximation of the solutions: increasing the order of the wavelet family and the increasing the resolution level of the wavelet. There is a growing interest in using various wavelets [1–3, 33–39] to study problems, of greater computational complexity. Among the wavelet transform families the Haar and Legendre wavelets deserve much attention. The basic idea of Legendre wavelet method is to convert the PDEs to a system of algebraic equations by the operational matrices of integral or derivative [40, 41]. The main goal is to show how wavelets and multi-resolution analysis can be applied for improving the method in terms of easy implementability and achieving the rapidity of its convergence. Razzaghi and Yousefi [40] introduced the Legendre wavelet method for solving variational problems and constrained optimal control problems. Hariharan et al. [1–3, 37] had introduced the diffusion equation, convection–diffusion equation, reaction–diffusion equation, nonlinear parabolic equations, fractional Klein–Gordon equations, sine-Gordon equations, and Fisher's equation by the Haar wavelet method. Mohammadi and Hosseini [35] had showed a new Legendre wavelet operational matrix of derivative in solving singular ordinary differential equations. Jafari et al. [42] had solved the fractional differential equations by the Legendre wavelet method. Parsian [41] introduced two-dimensional Legendre wavelets and operational matrices of integration. In recent years, many analytical/approximation methods have been proposed for solving Fisher's and fractional Fisher's equations. For example, Adomian decomposition method [5], the variational iteration method [16], the homotopy perturbation method [43], the differential transform method [17], the homotopy analysis method [21], and other methods [6]. Recently, Hariharan and Rajaraman [44] established a **new coupled wavelet-based method applied to the nonlinear reaction–diffusion equation arising in mathematical chemistry. Yin et al. [45] introduced a wavelet-based hybrid method for solving Klein–Gordon equations**.

In this work, we have applied a wavelet-based coupled method (LLWM) which combines the Laplace transform method and the Legendre wavelet method for the numerical solution of Fisher's and fractional Fisher's equations.

8.2 Method of Solution

8.2.1 Solving Fisher's and Fractional Fisher's Equations by the LLWM

We consider well-known Fisher's equation [2]

$$\frac{\partial U}{\partial t} = \frac{\partial^2 U}{\partial x^2} + \alpha U(1 - U) \tag{8.1}$$

with the initial condition

$$U(x,0) = f(x), \quad 0 \le x \le 1 \tag{8.2}$$

Taking Laplace transform on both sides of Eq. (8.1), we get

$$sL(U) - U(x,0) = L\big[U_{xx} + \alpha U - \alpha U^2\big] \tag{8.3}$$

$$sL(U) = U(x,0) + \big[L(U_{xx} + \alpha U - \alpha U^2)\big] \tag{8.4}$$

$$L(U) = \frac{U(x,0)}{s} + \frac{1}{s}L\big(U_{xx} + \alpha U - \alpha U^2\big) \tag{8.5}$$

Taking inverse Laplace transform to Eq. (8.5), we get

$$U(x,t) = U(x,0) + L^{-1}\left(\frac{1}{s}L\big(U_{xx} + \alpha U - \alpha U^2\big)\right) \tag{8.6}$$

Because

$$L^{-1}\left[\frac{1}{s}L(t^n)\right] = L^{-1}\left(\frac{n!}{s^{n+2}}\right) = \frac{1}{n+1}t^{n+1}; \quad (n = 0, 1, 2, \ldots) \tag{8.7}$$

We have

$$L^{-1}[s^{-1}L()] = \int_0^t (.)\mathrm{d}t \tag{8.8}$$

From Eq. (8.6)

$$U(x,t) = U(x,0) + L^{-1}\left(\frac{1}{s}L(U_{xx} + g(U))\right) \tag{8.9}$$

where $g(U) = \alpha U - \alpha U^2$

$$U(x,t) = U(x,0) + L^{-1}\left(\frac{1}{s}L(U_{xx} + g(U))\right) \tag{8.10}$$

By using the Legendre wavelet method,

$$\left.\begin{array}{l} U(x,t) = C^{\mathrm{T}}\psi(x,t) \\ U(x,0) = S^{\mathrm{T}}\psi(x,t) \\ g(U) = G^{\mathrm{T}}\psi(x,t) \end{array}\right\} \tag{8.11}$$

Substituting Eq. (8.11) in Eq. (8.6), we obtain

$$C^{\mathrm{T}} = S^{\mathrm{T}} + (C^{\mathrm{T}}Dx^2 - G^{\mathrm{T}})P_t^2 \tag{8.12}$$

Here G^{T} has a nonlinear relation with C. When we solve a nonlinear algebraic system, we get the solution is more complex and large computation time. In order to overcome the above drawbacks, we introduce an approximation formula as follows:

$$U_{n+1} = U(x,0) + \Pi\left[\frac{\partial^2 U_n}{\partial x^2} + g(U_n)\right] \tag{8.13}$$

where $g(U) = \alpha U - \alpha U^2$.

Expanding $u(x, t)$ by Legendre wavelets using the following relation

$$C_{n+1}^{\mathrm{T}} = C_0^{\mathrm{T}} + \left[C_n^{\mathrm{T}}D_x^2 - G_n^{\mathrm{T}}\right]P_t^2 \tag{8.14}$$

8.3   Convergence Analysis and Error Estimation [44, 45]

$$U^* = U_0 + \Pi\left[U_{xx}^* + g(U^*)\right] \tag{8.15}$$

$$U_{n+1} = U_0 + \Pi\left[(U_n)_{xx} + g(U_n)\right] \tag{8.16}$$

Subtracting Eq. (8.16) in Eq. (8.15), we obtain

$$U_{n+1} - U^* = \Pi\left[(U_n - U^*)_{xx} + g(U_n) - g(U^*)\right] \tag{8.17}$$

Using Lipschitz condition,

$\|g(U_n) - g(U^*)\| \le \gamma\|U_n - U^*\|$, we have
$$\|U_{n+1} - U^*\| \le \left\|\Pi(U_n - U^*)_{xx}\right\| + \left\|\Pi(g(U_n) - g(U^*))\right\| \tag{8.18}$$

$$\le \left\|\Pi(U_n - U^*)_{xx}\right\| + \gamma\left\|\Pi(U_n - U^*)\right\| \tag{8.19}$$

Let $U_{n+1} = C_{n+1}^{\mathrm{T}}\psi(x,t)$

$$U^* = C^{\mathrm{T}}\psi(x,t)$$

$$\epsilon_{n+1}^{\mathrm{T}} = C_{n+!}^{\mathrm{T}} - C^{\mathrm{T}}$$

Equation (8.5) gives

$$\epsilon_{n+1}^{\mathrm{T}} \leq \epsilon_n^{\mathrm{T}} \left\| D_x^2 P_t^2 + \gamma P_t^2 \right\| \tag{8.20}$$

Formula Eq. (8.7) can be obtained by using recursive relation.

$$\epsilon_{n+1}^{\mathrm{T}} \leq \epsilon_n^{\mathrm{T}} \left\| D_x^2 P_t^2 + \gamma P_t^2 \right\|^n \epsilon_0 \tag{8.21}$$

When $\underset{n \to \infty}{\mathrm{Lim}} \left\| D_x^2 P_t^2 + \gamma P_t^2 \right\| = 0$, the series solution of Eq. (8.1) using the LLWM converges to $u^*(x)$. By using the definitions of D_x and P_t, we can get the value of γ.
Suppose $k = k' = 1$ and $M = M'$, the maximum element of D_x and P_t is $2\sqrt{(2M-1)(2M-3)}$ and 0.5, respectively.

8.4 Illustrative Examples

Example 8.1 We consider Fisher's equation of the form [2]

$$\frac{\partial u}{\partial t} = \frac{\partial^2 u}{\partial x^2} + \alpha u(1-u) \tag{8.22}$$

Subject to the initial condition

$$u(x,0) = \frac{1}{\left(1 + e^{\sqrt{\frac{7}{6}}x}\right)^2} \tag{8.23}$$

Using homotopy analysis method (HAM), the exact solution in a closed form is given by

$$u(x,t) = \frac{1}{\left(1 + e^{\sqrt{\frac{7}{6}}x - \frac{5}{6}at}\right)^2} \tag{8.24}$$

The Haar wavelet scheme (HWS) of Eq. (8.1) is given by

$$c_{(m)}^{\mathrm{T}} Q_{(m)} h_{(m)}(x_l) + x_l \left[-c_{(m)}^{\mathrm{T}} P_{(m)} \lambda + g_1'(t_{s+1}) - g_0'(t_{s+1}) \right] + g_0'(t_{s+1})$$
$$= u''(x_l, t_{s+1}) + \alpha u(x_l, t_{s+1})[1 - u(x_l, t_{s+1})]$$

From the above formula, the wavelet coefficients $c_{(m)}^{\mathrm{T}}$ can be successively calculated.

Our proposed method (LLWM) can be compared with Wazwaz and Gorguis results (see Ref. [5]), Merdan results (see Ref. [30]), Hariharan and Kannan [2], and Zhou [46] results. Good agreement with the exact solution is observed.

Example 8.2 Consider the Fisher equation of the form [2, 5, 18]

$$\frac{\partial u}{\partial t} = \frac{\partial^2 u}{\partial x^2} + u^2(1-u), \quad 0 < x < 1 \tag{8.25}$$

With the initial condition

$$u(x,0) = \frac{1}{1+e^{\frac{x}{\sqrt{2}}}} \tag{8.26}$$

Using the HAM, the exact solution in a closed form is given by

$$u(x,t) = \frac{1}{1+e^{v(x-vt)}}, \quad v = \frac{1}{\sqrt{2}} \tag{8.27}$$

Our proposed method (LLWM) can be compared with Wazwaz and Gorguis results (see Ref. [5]) and Mehmet Merdan results (see Ref. [30]), Hariharan et al. [2], and Zhou [46] results. Good agreement with the exact solution is observed.

Example 8.3 Let us consider the following time-fractional Fisher's reaction–diffusion equation

$$D_t^\alpha u = u_{xx} + 6u(1-u), \quad t > 0, \ x \in R \tag{8.28}$$

With initial condition

$$u(x,0) = \frac{1}{(1+e^x)^2} \tag{8.29}$$

Using differential transform method (DTM), the series solution is given by

$$\begin{aligned}
u(x,t) &= \frac{1}{4} - \frac{1}{4}x + \frac{1}{16}x^2 + \frac{1}{48}x^3 + \left(\frac{5}{4} - \frac{5}{8}x - \frac{5}{16}x^2\right)\frac{t^\alpha}{\Gamma(\alpha+1)} \\
&\quad + \left(\frac{25}{16} + \frac{25}{16}x\right)\frac{t^{2\alpha}}{\Gamma(2\alpha+1)} - \frac{125}{8}\frac{t^{3\alpha}}{\Gamma(3\alpha+1)} + \cdots
\end{aligned} \tag{8.30}$$

When $\alpha = 1$, the exact solution is given by

$$u(x,t) = \frac{1}{(1+e^{x-5t})^2} \tag{8.31}$$

Tables 8.1, 8.2, 8.3, and 8.4 show the numerical solutions of Fisher's equations and fractional Fisher's equations for various values (x, t) and $\alpha = 1$. Our LLWM results are in excellent agreement with the exact solution, the homotopy analysis method (HAM) and the differential transform method (DTM). Figures 8.1, 8.2, 8.3, 8.4, 8.5,

Table 8.1 Comparison between the exact and LLWM for Example 8.1

x	t	U_{exact}	U_{LLWM}
0.25	0.5	0.81839	0.81855
	1.0	0.98292	0.98305
	2.0	0.99988	0.99999
	5.0	1.00000	1.00000
0.50	0.5	0.77590	0.77602
	1.0	0.97815	0.97824
	2.0	0.99985	0.99996
	5.0	1.00000	1.0000
0.75	0.5	0.72582	0.72595
	1.0	0.92207	0.92221
	2.0	0.99981	0.99993
	5.0	1.00000	1.00000

Table 8.2 Comparison between the exact and LLWM for Example 8.2

x	t	U_{exact}	U_{LLWM}
0.25	0.5	0.51830	0.51839
	1.0	0.58011	0.58018
	2.0	0.69492	0.69599
	5.0	0.91078	0.91085
	8.5	0.98331	0.98336
	11.0	0.99513	0.99514
0.50	0.5	0.47414	0.47423
	1.0	0.53655	0.53661
	2.0	0.65621	0.65626
	5.0	0.89533	0.89535
	8.5	0.98012	0.98015
	11.0	0.99423	0.99424
0.75	0.5	0.43037	0.43047
	1.0	0.49242	0.49252
	2.0	0.61531	0.61539
	5.0	0.87757	0.87765
	8.5	0.97633	0.97636
	11.0	0.99312	0.99314

and 8.6 show the numerical solutions of Fisher's equation and fractional Fisher's equations for various values of (x, t) and $\alpha = 1$.

All the numerical experiments presented in this section were computed in double precision with some MATLAB codes on a personal computer System Vostro 1400 Processor x86 Family 6 Model 15 Stepping 13 Genuine Intel ~ 1596 MHz.

Table 8.3 Comparison between the exact and LLWM for Example 8.3

x	t	U_{exact}	U_{LLWM}
0.25	0.5	0.8184	0.8186
	1.0	0.9829	0.9832
	1.5	0.9999	0.9999
	2.0	1.0000	1.0000
0.50	0.5	0.7758	0.7761
	1.0	0.9781	0.9783
	1.5	0.9999	1.0000
	2.0	1.0000	1.0000
0.75	0.5	0.7258	0.7261
	1.0	0.9721	0.9723
	1.5	0.9998	0.9999
	2.0	1.0000	1.0000

Table 8.4 Comparison between exact solution and LLWM for Example 8.3 for different values of x and t

x	t	Exact	LLWM
0.1	0.2	0.5054	0.5062
0.2	0.4	0.7364	0.7371
0.3	0.6	0.8780	0.8786
0.4	0.8	0.9475	0.9480
0.5	1.0	0.9781	0.9784
0.6	1.2	0.9910	0.9913
0.7	1.4	0.9963	0.9966
0.8	1.6	0.9985	0.9986
0.9	1.8	0.9994	0.9994
1.0	2.0	0.9998	0.9998

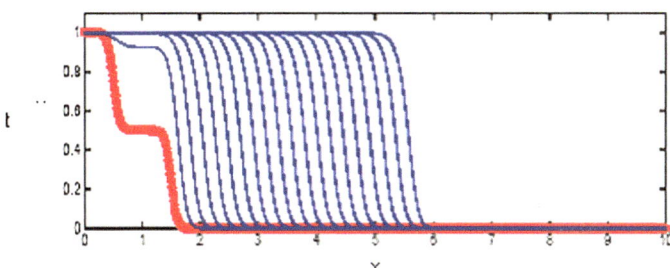

Fig. 8.1 Numerical solutions of Fisher's equation for (x, t) and $\alpha = 0.5, k = 1$ and $M = 4$

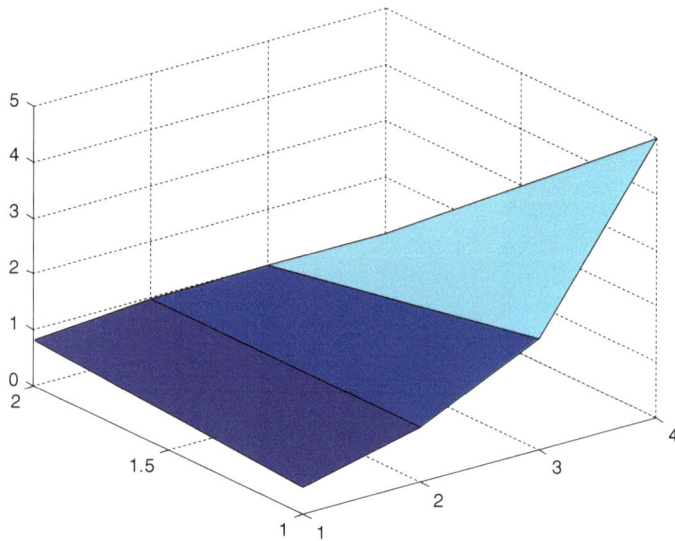

Fig. 8.2 Surface area shows that $u(x, t)$ using LLWM for Eq. (8.1) at $x = 0.25$, $k = 1$ and $M = 4$

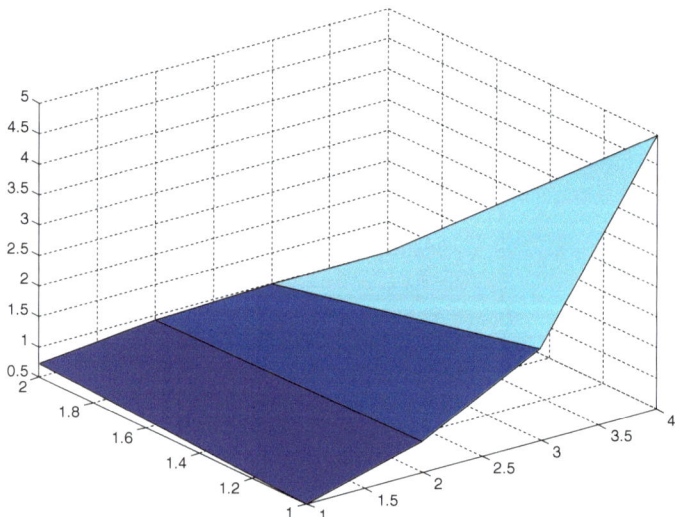

Fig. 8.3 Surface area shows that $u(x, t)$ using LLWM for Eq. (8.1) at $x = 0.75$, $k = 1$ and $M = 4$

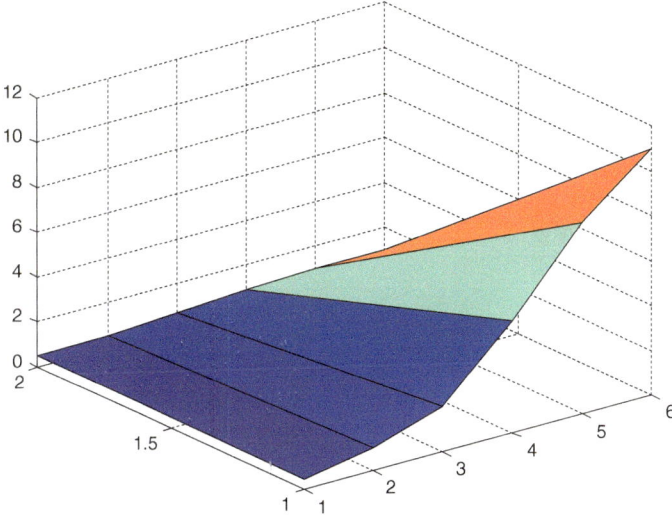

Fig. 8.4   Surface area shows that $u(x, t)$ using LLWM for Eq. (8.2) at $x = 0.25$, $k = 1$ and $M = 4$

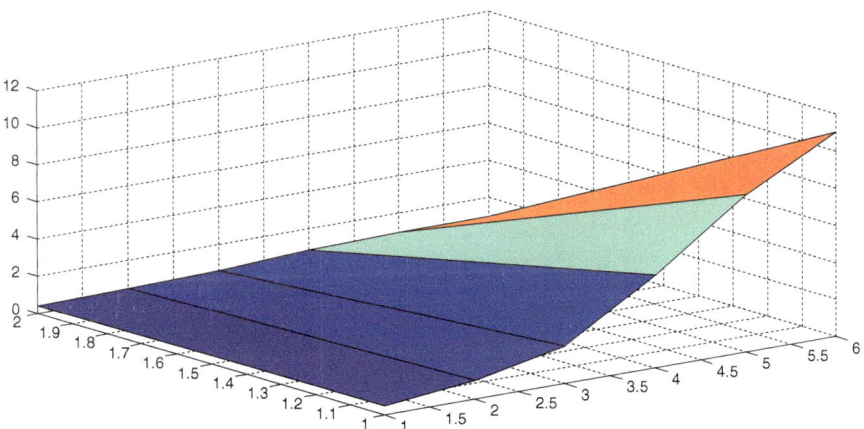

Fig. 8.5   Surface area shows that $u(x, t)$ using LLWM for Eq. (8.2) at $x = 0.75$, $k = 1$ and $M = 4$

8.5   Conclusion

In this work, a new coupled wavelet-based method has been successfully employed to obtain the numerical solutions of Fisher's and time-fractional Fisher's equations arising in population genetics. The proposed scheme is the capability to overcome the difficulty arising in calculating the integral values while dealing with nonlinear problems. This method shows higher efficiency than the traditional Legendre

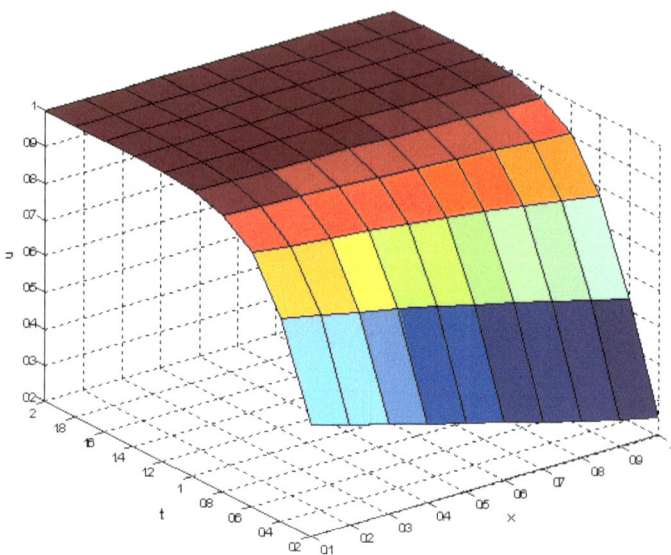

Fig. 8.6 Surface area shows that $u(x, t)$ using LLWM for Eq. (8.3) for various values of (x, t) and $M = 4$

wavelet method for solving nonlinear PDEs. Numerical example illustrates the powerful of the proposed scheme LLWM. Also, this paper illustrates the validity and excellent potential of the LLWM for nonlinear and fractional PDEs. The numerical solutions obtained using the proposed method show that the solutions are in very good coincidence with the exact solution. In addition, the calculations involved in LLWM are simple, straightforward and low computation cost. In Sect. 8.4, we have developed the convergence of the proposed algorithm.

Appendix: Basic Idea of Homotopy Analysis Method (HAM)

In this section, the basic ideas of the homotopy analysis method are presented. Here a description of the method is given to handle the general nonlinear problem.

$$Nu_0(t) = 0, \quad t > 0 \tag{8.32}$$

Where N is a nonlinear operator and $\boldsymbol{u}_0(t)$ is unknown function of the independent variable t.

Zero-order deformation equation

Let $u_0(t)$ denote the initial guess of the exact solution of Eq. (8.1), h $\neq 0$ an auxiliary parameter, $H(t) \neq 0$ an auxiliary function and L is an auxiliary linear operator with the property.

$$L(f(t)) = 0, \quad f(t) = 0. \tag{8.33}$$

The auxiliary parameter h, the auxiliary function $H(t)$, and the auxiliary linear operator L play an important role within the HAM to adjust and control the convergence region of solution series. Liao [30] constructs, using $q \in [0, 1]$ as an embedding parameter, the so-called zero-order deformation equation.

$$(1 - q)L[(\emptyset(t; q) - u_0(t)] = qhH(t)N[(\emptyset(t; q)], \tag{8.34}$$

where $\emptyset(t; q)$ is the solution which depends on h, $H(t), L, u_0(t)$ and q. When $q = 0$, the zero-order deformation Eq. (8.33) becomes

$$\emptyset(t; 0) = u_0(t), \tag{8.35}$$

And when $q = 1$, since $h \neq 0$ and $H(t) \neq 0$, the zero-order deformation Eq. (8.32) reduces to,

$$N[\emptyset(t; 1)] = 0, \tag{8.36}$$

So, $\emptyset(t; 1)$ is exactly the solution of the nonlinear equation. Define the so-called mth-order deformation derivatives.

$$u_m(t) = \frac{1}{m!} \frac{\partial^m \emptyset(t; q)}{\partial q^m} \tag{8.37}$$

If the power series Eq. (8.34) of $\emptyset(t; q)$ converges at $q = 1$, then we gets the following series solution:

$$u(t) = u_0(t) + \sum_{m=1}^{\infty} u_m(t). \tag{8.38}$$

where the terms $u_m(t)$ can be determined by the so-called high-order deformation described below.

High-order deformation equation

Define the vector,

$$\overrightarrow{u_n} = \{u_0(t), u_1(t), u_2(t)\ldots u_n(t)\} \tag{8.39}$$

Differentiating Eq. (8.34) m times with respect to embedding parameter q, the setting $q = 0$ and dividing them by !, we have the so-called mth-order deformation equation.

$$L[u_m(t) - \aleph_m u_{m-1}(t)] = hH(t)R_m(\overrightarrow{u_m}, t), \tag{8.40}$$

where

$$\aleph_m = \begin{cases} 0, & m \leq 1 \\ 1, & \text{otherwise} \end{cases} \tag{8.41}$$

and

$$R_m(\overrightarrow{u_m}, t) = \frac{1}{(m-1)!} \frac{\partial^{m-1} N[\emptyset(t; q)]}{\partial q^{m-1}} \tag{8.42}$$

For any given nonlinear operator N, the term $R_m(\overrightarrow{u_m}, t)$ can be easily expressed by Eq. (8.42). Thus, we can gain $u_1(t), u_2(t) \ldots$ by means of solving the linear high-order deformation with one after the other order in order. The mth-order approximation of $u(t)$ is given by

$$u(t) = \sum_{k=0}^{m} u_k(t) \tag{8.43}$$

ADM, VIM, and HPM are special cases of HAM when we set $h = -1$ and $H(r,t) = 1$. We will get the same solutions for all the problems by above methods when we set $h = -1$ and $H(r,t) = 1$. When the base functions are introduced, the $H(r,t) = 1$ is properly chosen using the rule of solution expression, rule of coefficient of ergodicity, and rule of solution existence.

References

1. G. Hariharan, K. Kannan, K. Sharma, Haar wavelet in estimating the depth profile of soil temperature. Appl. Math. Comput. **210**, 119–225 (2009)
2. G. Hariharan, K. Kannan, Haar wavelet method for solving Fisher's equation. Appl. Math. Comput. **211**, 284–292 (2009)
3. G. Hariharan, K. Kannan, Haar wavelet method for solving nonlinear parabolic equations. J. Math. Chem. **48**, 1044–1061 (2010)
4. J.D. Murray, *Lectures on non-linear differential equation models in biology* (Clarenden, Oxford, 1977)
5. A.M. Wazwaz, A. Gorguis, An analytical study of Fisher's equation by using Adomian decomposition method. Appl. Math. Comput. **154**, 609–620 (2004)
6. D. Olmos, B. Shizgal, A spectral method of solution of Fisher's equation. J. Comput. Appl. Math. **193**, 219–242 (2006)

7. V. Turut, N. Guzel, Comparing numerical methods for solving time-fractional reaction-diffusion equations. ISRN Math. Anal. **2012**, Article ID 737206 (2012). https://doi.org/10.5402/2012/737206

8. F.C. Meral, T.J. Royston, R. Magin, Fractional calculus in viscoelasticity: an experimental study. Commun. Nonlinear Sci. Numer. Simul. **15**(4), 939–945 (2010)

9. K. Seki, M. Wojcik, M. Tachiya, Fractional reaction–diffusion equation. J. Chem. Phys. **119**, 2165–2174 (2003)

10. B.I. Henry, S.L. Wearne, Fractional reaction-diffusion. Physica A **276**(3–4), 448–455 (2000)

11. B. Baeumer, M. Kov´acs, M.M. Meerschaert, Numerical solutions for fractional reaction-diffusion equations. Comput. Math. Appl. **55**(10), 2212–2226 (2008)

12. S.Z. Rida, A.M.A. El-Sayed, A.A.M. Arafa, On the solutions of time-fractional reaction-diffusion equations. Commun. Nonlinear Sci. Numer. Simul. **5**(12), 3847–3854 (2010)

13. S. Momani, R. Qaralleh, Numerical approximations and Padé approximants for a fractional population growth model. Appl. Math. Model. **31**(9), 1907–1914 (2007)

14. A. Cuyt, L. Wuytack, *Nonlinear Methods in Numerical Analysis* (Elsevier Science, Amsterdam, 1987)

15. H.A. Abdusalam, Analytic and approximate solutions for Nagumo telegraph reaction diffusion equation. Appl. Math. Comput. **157**, 515–522 (2004)

16. M. Matinfar, M. Ghanbari, Solving the Fisher's equation by means of variational iteration method. Int. J. Contemp. Math. Sci. **4**(7), 343–348 (2009)

17. M. Matinfar, S.R. Bahar, M. Ghasemi, Solving the generalized Fisher's equation by the differential transform method. J. Appl. Math. Inf. **30**(3–4), 555–560 (2012)

18. K. Yıldırım, B. İbiş, M. Bayram, New solutions of the nonlinear Fisher type equations by the reduced differential transform. Nonlinear Sci. Lett. A **3**(1), 29–36 (2012)

19. N.A. Khan, N.-U. Khan, A. Ara, M. Jamil, Approximate analytical solutions of fractional reaction-diffusion equations. J. King Saud Univ. Sci. 24, 111–118 (2012)

20. S.J. Liao, *Beyond perturbation: introduction to homotopy analysis method* (CRC Press/Chapman and Hall, Boca Raton, 2004)

21. G. Hariharan, The homotopy analysis method applied to the Kolmogorov–Petrovskii–Piskunov (KPP) and fractional KPP equations. J. Math. Chem. **51**, 992–1000 (2013). https://doi.org/10.1007/s10910-012-0132-5

22. J.H. He, X.H. Wu, Exp-function method for nonlinear wave equations. Chaos, Solitons Fractals **30**, 700–708 (2006)

23. X.J. Yang, *Local fractional functional analysis and its applications* (Asian Academic, Hong Kong, 2011)

24. X.J. Yang, Local fractional integral transforms. Prog. Nonlinear Sci. **4**, 1–225 (2011)

25. X.J. Yang, *Advanced local fractional calculus and its applications* (World Science, New York, 2012)

26. X.J. Yang, D. Baleanu, Fractal heat conduction problem solved by local fractional variation iteration method. Therm. Sci. **17**(2), 625–628 (2013)

27. G.F. Carey, Y. Shen, Least-squares finite element approximation of Fisher's reaction–diffusion equation. Numer. Methods Partial Differ. Equ. **11**(2), 175–186 (1995)

28. K. Al-Khaled, Numerical study of Fisher's reaction-diffusion equation by the Sinc collocation method. J. Comput. Appl. Math. **137**, 245–255 (2001)

29. R.C. Mittal, R. Jiwari, Numerical strudy of Fisher's equation by using differential quadrature method. Int. J. Inf. Syst. Sci. **5**(1), 143–160 (2008)

30. M. Merdan, Solutions of time-fractional reaction-diffusion equation with modified Riemann-Liouville derivative. Int. J. Phys. Sci. **7**(15), 2317–2326 (2012)

31. M. Kurulay, M. Bayram, Comparison of numerical solutions of time-fractional reaction-diffusion equations. Malaysian J. Math. Sci. **6**(S), 49–59 (2012)

32. A-M. Yang, Y-Z. Zhang, Y. Long, The Yang-Fourier transforms to heat-conduction in a semi-infinite fractal bar. Therm. Sci. **17**(3), 707–713 (2013)

33. M. Razzaghi, S. Yousefi, The Legendre wavelets direct method for variational problems. Math. Comput. Simulat. **53**, 185–192 (2000)
34. S.A. Yousefi, Legendre wavelets method for solving differential equations of Lane-Emden type. App. Math. Comput. **181**, 1417–1442 (2006)
35. F. Mohammadi, M.M. Hosseini, A new Legendre wavelet operational matrix of derivative and its applications in solving the singular ordinary differential equations. J. Franklin Inst. **348**, 1787–1796 (2011)
36. K. Maleknejad, S. Sohrabi, Numerical solution of Fredholm integral equations of the first kind by using Legendre wavelets. Appl. Math. Comput. **186**, 836–843 (2007)
37. G. Hariharan, K. Kannan, A comparative study of a haar wavelet method and a restrictive Taylor's series method for solving convection-diffusion equations. Int. J. Comput. Methods Eng. Sci. Mech. **11**(4), 173–184 (2010)
38. Yin Yang, Solving a nonlinear multi-order fractional differential equation using legendre pseudo-spectral method. Appl. Math. **4**, 113–118 (2013). https://doi.org/10.4236/am.2013.41020
39. M.H. Heydari, M.R. Hooshmandasl, F.M. Maalek Ghaini, F. Mohammadi, Wavelet collocation method for solving multiorder fractional differential equations. J. Appl. Math. **2012**, Article ID 54240 (2012). https://doi.org/10.1155/2012/542401
40. M. Razzaghi, S. Yousefi, The Legendre wavelets operational matrix of integration. Int. J. Syst. Sci. **32**, 495–502 (2001)
41. H. Parsian, Two dimension Legendre wavelets and operational matrices of integration. Acta Math. Acad. Paedagog Nyhaza. **21**, 101–106 (2005)
42. H. Jafari, M. Soleymanivaraki M.A. Firoozjaee, Legendre wavelets for solving fractional differential equations. J. Appl. Math. **7**(427), 65–70 (2011)
43. M. Matinfar, M. Ghanbari, Homotopy perturbation method for the Fisher's equation and its generalized. Int. J. Nonlinear Sci. **8**(4), 448–455 (2009)
44. G. Hariharan, R. Rajaraman, A new coupled wavelet-based method applied to the nonlinear reaction–diffusion equation arising in mathematical chemistry. J. Math. Chem. **51**, 2386–2400 (2013). https://doi.org/10.1007/s10910-013-0217-9
45. F. Yin, J. Song, F. Lu, A coupled method of Laplace transform and Legendre wavelets for nonlinear Klein–Gordon equations. Meth. Methods Appl. Sci. (2013). https://doi.org/10.1002/mma.2834
46. X.W. Zhou, Exp-function method for solving Fisher's equation. J. Phys: Conf. Ser. **96**, 012063 (2008)

Chapter 9
Two Reliable Wavelet Methods to Fitzhugh–Nagumo (FN) and Fractional FN Equations

Fractional reaction–diffusion equations (FRDEs) serve as more relevant models for studying complex patterns in several fields of nonlinear sciences. In this paper, we have developed the wavelet methods to find the approximate solutions for the Fitzhugh–Nagumo (FN) and fractional FN equations. The proposed method techniques provide the solutions in rapid convergence series with computable terms. To the best of our knowledge, until now, there is no rigorous wavelet solutions have been reported for the Fitzhugh–Nagumo (FN) and fractional FN equations arising in gene propagation and model. With the help of Laplace operator and Legendre wavelets operational matrices, the Fitzhugh–Nagumo (FN) equation is converted into an algebraic system. Finally, we have given some numerical examples to demonstrate the validity and applicability of the wavelet methods. The power of the manageable method is confirmed. Moreover, the use of the wavelet methods is found to be accurate, efficient, simple, low computation costs, and computationally attractive.

9.1 Introduction

In recent years, nonlinear reaction–diffusion equations (NLRDEs) have been widely studied and applied in science, engineering, and medicine [1, 2]. Reaction–diffusion equations (RDEs) are commonly applied to model the growth and spreading of biological species [3]. A fractional reaction–diffusion equation (FRDE) can be derived from a continuous-time random walk model when the transport is dispersive or a continuous-time random walk model with temporal memory and sources [4]. In recent years, the FRDE has received the applications in systems biology [5–7], chemistry, and biochemistry applications [8]. Another time-fractional reaction–diffusion equation is the time-fractional Fitzhugh–Nagumo (FN) equation. It is an important nonlinear reaction–diffusion equation in population genetics [9], circuit

© Springer Nature Singapore Pte Ltd. 2019

G. Hariharan, *Wavelet Solutions for Reaction–Diffusion Problems in Science and Engineering*, Forum for Interdisciplinary Mathematics, https://doi.org/10.1007/978-981-32-9960-3_9

theory [6, 10, 11], ventricle tissue model [9, 12–14] and usually used to model the transmission of nerve impulses [9, 10].

In 1952, Hodgkin and Huxley [15] developed an efficient FN models for the conduction of nerve impulses along axon. They established a mathematical model to describe the membrane's behavior by considering the conduction and excitation of the fiber [16]. The Fitzhugh–Nagumo (FN) models were derived by both Fitzhugh [10] and Nagumo et al. [17]. In recent years, these models are important nonlinear reaction–diffusion models used in circuit theory, biology, and the area of population genetics [9]. The FN model equations describe the dynamical behavior near the bifurcation point for the Rayleigh–Benard convection of binary fluid mixtures [18]. But these nonlinear PDEs are difficult to get their exact solutions. So, the approximation and numerical methods must be used. The numerical solutions of the nonlinear reaction–diffusion equations (NLRDEs) have received considerable attention in the literature and fall into two groups: the analytical methods and the numerical ones. Analytical methods enable researchers to study the effect of different variables or parameters on the function under study easily. Recently, there are many new algorithms for NLRDEs have been proposed, for example, the Adomian decomposition method [19, 20], pseudo-spectral method [21, 22], generalized differential transform method [23], the homotopy analysis method [7, 13, 24, 25], Haar wavelet method [19, 26–31], Legendre wavelet method [32–34], and other methods [35–38]. Recently, Khan et al. [13] introduced the approximate analytical solutions of fractional reaction–diffusion equations.

Wavelets theory is a relatively new and as emerging area in applied mathematical research. It has been applied in many different fields of science and engineering. Moreover, the wavelet transform establishes a connection with efficient and fast numerical algorithms.

In recent years, wavelet transforms have found their way into many different fields in science, engineering, and medicine. Wavelet analysis or wavelet theory, as a relatively new and emerging area in applied mathematical research, has received considerable attention in dealing with nonlinear reaction–diffusion equations (NLRDEs). It possesses many useful properties, such as compact support, orthogonality, dyadic, orthonormality, and multiresolution analysis (MRA) [19, 26–28]. Recently, Haar wavelets have been applied extensively for signal processing in communications and physics research and have proved to be a wonderful mathematical tool. After discretizing the differential equations in a conventional way like the finite difference approximation, wavelets can be used for algebraic manipulations in the system of equations obtained which lead to the better condition number of the resulting system.

In the numerical analysis, wavelet-based methods and hybrid methods become important tools because of the properties of localization. In wavelet-based methods, there are two important ways of improving the approximation of the solutions: increasing the order of the wavelet family and increasing the resolution level of the wavelet. There is a growing interest in using various wavelets [19, 26–28, 32–34] to study problems, of greater computational complexity. Among the wavelet transform families, the Haar and Legendre wavelets deserve much attention. The basic idea of Legendre wavelet method is to convert the partial differential equations to a system of algebraic equations by the operational matrices of integral or derivative [28]. The main goal is to show how wavelets and multiresolution analysis can be applied for improving the method in terms of easy implementability and achieving the rapidity of its convergence. Razzaghi and Yousefi [39, 40] introduced the Legendre wavelet method for solving variational problems and constrained optimal control problems. Recently, Hariharan et al. [19, 26–28] introduced the diffusion equation, convection–diffusion equation, reaction–diffusion equation, nonlinear parabolic equations, fractional Klein–Gordon equations, Sine-Gordon equations, and Fisher's equation by the Haar wavelet method. Mohammadi and Hosseini [41] had shown a new Legendre wavelet operational matrix of derivative in solving singular ordinary differential equations. Parsian [42] introduced two-dimensional Legendre wavelets and operational matrices of integration. Yousefi [43] introduced the Legendre wavelets for solving Lane–Emden-type differential equations. Recently, Fukang Yin et al. [44] introduced the a coupled method of Laplace transform and Legendre wavelets for Lane–Emden-type differential equations,

In this chapter, we have applied Haar and Legendre wavelets for Fitzhugh–Nagumo (FN) and time-fractional FN equations arising in population genetics.

9.2 Method of Solution

9.2.1 Solving Fitzhugh–Nagumo (FN) Equation by the Haar Wavelet Method (HWM)

We consider the Fitzhugh–Nagumo (FN) equation [29]

$$\frac{\partial u}{\partial t} = \frac{\partial^2 u}{\partial x^2} + u(u - \alpha)(1 - u) \tag{9.1}$$

Since $u(x, t) \in L^2(R)$, we suppose

$$u(x, t) \approx \sum_{i=0}^{m-1} \sum_{j=0}^{m-1} c_{ij} h_i(x) h_j(t), \tag{9.2}$$

Then, we can obtain the discrete form of Eq. (9.2) by taking step $\Delta = 1/m$ of x, t; there are

$$u = H^{\mathrm{T}}(x)CH(t) \tag{9.3}$$

$$\frac{\partial u}{\partial t} \approx H^{\mathrm{T}}(x)C\frac{\partial}{\partial t}H(t)$$
$$= H^{\mathrm{T}}(x)CQ_H^{-1}H(t) \tag{9.4}$$

$$\frac{\partial^2 u}{\partial x^2} \approx H^{\mathrm{T}}(x)\left(Q_H^{-2}\right)^{\mathrm{T}}CH(t) \tag{9.5}$$

Substitute Eqs. (9.3)–(9.5) into Eq. (9.1), there is

$$
\begin{aligned}
H^{\mathrm{T}}(x)CQ_H^{-1}H(t) = {} & H^{\mathrm{T}}(x)\left(Q_H^{-2}\right)^{\mathrm{T}}CH(t) \\
& + H^{\mathrm{T}}(x)CH(t)\left(H^{\mathrm{T}}(x)CH(t) - \alpha\right) \\
& \times \left(1 - H^{\mathrm{T}}(x)CH(t)\right)
\end{aligned}
\tag{9.6}
$$

From the above formula, the wavelet coefficient C can be calculated successively.

9.3   Solving Fitzhugh–Nagumo (FN) Equation by the LLWM

We consider the Eq. (9.1) [29]

Taking Laplace transform on both sides of Eq. (9.1), we get

$$sL(u) - u(x,0) = L\left[u_{xx} - u^2 - u^3 - u\alpha + u^2\alpha\right] \tag{9.7}$$

$$sL(u) = u(x,0) + L\left[u_{xx} - u^2 - u^3 - u\alpha + u^2\alpha\right] \tag{9.8}$$

$$L(u) = s^{-1}(u(x,0)) + s^{-1}\left(L\left[u_{xx} - u^2 - u^3 - u\alpha + u^2\alpha\right]\right) \tag{9.9}$$

Taking inverse Laplace transform on both sides of Eq. (9.9)

$$u = u(x,0) + L^{-1}\left(s^{-1}L\left[u_{xx} - u^2 - u^3 - u\alpha + u^2\alpha\right]\right) \tag{9.10}$$

Because

$$
\begin{aligned}
L^{-1}[s^{-1}(t^n)] &= L^{-1}(n!s^{-(n+2)}) \\
&= \frac{1}{n+1}t^{n+1}; \quad (n = 0, 1, 2, \ldots)
\end{aligned}
\tag{9.11}
$$

We have

$$L^{-1}\left[s^{-1}L(\cdot)\right] = \int_0^t (\cdot)\,dt \tag{9.12}$$

From Eq. (9.10), we gain

$$u = u(x,0) + L^{-1}\left(s^{-1}L[u_{xx} + g(u)]\right), \tag{9.13}$$

where $g(u) = u^2\alpha - u\alpha + u^2 - u^3$

Using the Legendre wavelets method,

$$u = C^{\mathrm{T}}\psi(x,t) \tag{9.14}$$

$$\left.\begin{array}{l} u(x,0) = S^{\mathrm{T}}\psi(x,t) \\ g(u) = G^{\mathrm{T}}\psi(x,t) \end{array}\right\} \tag{9.15}$$

Substituting Eqs. (9.14) and (9.15) in Eq. (9.13), we obtain

$$C^{\mathrm{T}} = S^{\mathrm{T}} + \left(C^{\mathrm{T}}Dx^2 + G^{\mathrm{T}}\right)P_t^2 \tag{9.16}$$

Here, G^{T} has a nonlinear relation with C. When we solve a nonlinear algebraic system, we get the solution is more complex and large computation time. In order to overcome the above drawbacks, we introduce an approximation formula as follows:

$$u_{n+1} = u(x,0) + \Pi\left(\frac{\partial^2 u_n}{\partial x^2} + g(u_n)\right), \tag{9.17}$$

where $g(u) = u^2\alpha - u\alpha + u^2 - u^3$.

We start the first iteration; an initial guess of the solution of u_0 is required. We select

$u_0 = u(x,0)$, and expanding u by Legendre wavelets, we gain

$$C_{n+1}^{\mathrm{T}} = C_0^{\mathrm{T}} + \left[C_n^{\mathrm{T}}D_x^2 + G_n^{\mathrm{T}}\right]P_t^2 \tag{9.18}$$

9.4 Convergence Analysis

$$u^* = u_0 + \prod \left(u_{xx}^* + g(u^*) \right) \tag{9.19}$$

and

$$u_{n+1} = u_0 + \prod \left((u_n)_{xx} + g(u_n) \right) \tag{9.20}$$

Subtracting Eq. (9.20) from Eq. (9.19), we obtain

$$u_{n+1} - u^* = \Pi \left[(u_n - u^*)_{xx} + (g(u_n) - g(u^*)) \right] \tag{9.21}$$

Using Lispschitz condition,

$$\| g(u_n) - g(u^*) \| \leq \gamma \| u_n - u^* \| \tag{9.22}$$

We have

$$\| u_{n+1} - u^* \| \leq \left\| \prod (u_n - u^*)_{xx} \right\| + \left\| \prod (g(u_n) - g(u^*)) \right\| \tag{9.23}$$

$$\leq \left\| \prod (u_n - u^*)_{xx} \right\| + \gamma \| u_n - u^* \| \tag{9.24}$$

Let

$$u_{n+1} = C_{n+1}^{\mathrm{T}} \psi(x, t) \tag{9.25}$$

$$u^* = C^{\mathrm{T}} \psi(x, t) \tag{9.26}$$

$$\in_{n+1}^{\mathrm{T}} = C_{n+1}^{\mathrm{T}} - C^{\mathrm{T}} \tag{9.27}$$

From Eq. (9.27), we obtain the formula

$$\in_{n+1}^{\mathrm{T}} \leq \in_n^{\mathrm{T}} \left\| D_x^2 P_t^2 + \gamma P_t^2 \right\| \tag{9.28}$$

By recursion, we get

$$\in_{n+1}^{\mathrm{T}} \leq \in_n^{\mathrm{T}} \left\| D_x^2 P_t^2 + \gamma P_t^2 \right\|^n \in_0 \tag{9.29}$$

When $\underset{n \to \infty}{\mathrm{Lim}} \left\| D_x^2 P_t^2 + \gamma P_t^2 \right\|^n = 0$, the series solution of Eq. (9.1) using the LLWM converges to $u^*(x)$. By using the definitions of D_x and P_t, we can get the value of γ. Suppose $k = k' = 1$ and $M = M'$, the maximum element of D_x and P_t is

$2\sqrt{(2M - 1)(2M - 3)}$ and 0.5, respectively.

9.5   Numerical Examples

In this section, three examples are given for demonstrating the validity and applicability of the proposed wavelet methods.

Example 9.1 Consider the Fitzhugh–Nagumo (FN) equation

$$\frac{\partial u}{\partial t} = \frac{\partial^2 u}{\partial x^2} + u(u - \alpha)(1 - u), \quad 0 < \alpha < 1 \tag{9.30}$$

Subject to the initial condition

$$u(x, 0) = f(x) \tag{9.31}$$

Using the homotopy analysis method (HAM), the exact solution in a closed form is given by

$$u(x, t) = \frac{1}{1 + e^{\left(\frac{-x + ct}{\sqrt{2}}\right)}}, \tag{9.32}$$

which is full agreement with the results in [20] where $c = \sqrt{2}\left(\frac{1}{2} - \alpha\right)$.

The Haar wavelet scheme is given by

$$H^{\mathrm{T}}(x)CQ_H^{-1}H(t) = H^{\mathrm{T}}(x)\left(Q_H^{-2}\right)^{\mathrm{T}}CH(t)$$
$$+ H^{\mathrm{T}}(x)CH(t)\left(H^{\mathrm{T}}(x)CH(t) - \alpha\right)\left(1 - H^{\mathrm{T}}(x)CH(t)\right)$$

Our proposed wavelet methods HWM and LLWM can be compared with Wazwaz and Gorguis results (see Ref. [20]) and Mehmet Merdan results (see Ref. [45]).

Example 9.2 We consider the time-fractional Fitzhugh–Nagumo (FN) equation [20]

$$\frac{\partial^\alpha u}{\partial t^\alpha} = \frac{\partial^2 u}{\partial x^2} + u(u - \mu)(1 - u), \quad \mu > 0, \ 0 < \alpha \le 1, \ t > 0, \ x \in \Re \tag{9.33}$$

Subject to the initial condition

$$u(x, 0) = \frac{1}{1 + e^{\left(\frac{-x}{\sqrt{2}}\right)}} \tag{9.34}$$

As $\alpha \to 1$ and $h = -1$, the exact solution of Eq. (9.4) in a closed form by the homotopy analysis method (HAM) is given by

Table 9.1 Numerical values when $\alpha = 0.5$ and $\mu = 0.45$ for Example 9.2

x	t	Exact solution u_{HAM}	Numerical u_{HWM} ($m = 32$)	Numerical u_{LLWM} ($k = 1$ and $M = 3$)
0.001	0.001	0.50026800	0.50026801	0.50026802
0.002	0.002	0.50027473	0.50027473	0.50027471
0.003	0.003	0.50017995	0.50017993	0.50017993
0.004	0.004	0.50010722	0.50010721	0.50010722
0.005	0.005	0.50002438	0.50002435	0.50002435
0.006	0.006	0.49993390	0.49993390	0.49993391
0.007	0.007	0.49998373	0.49998374	0.49998372
0.008	0.008	0.49992356	0.49992357	0.49992353
0.009	0.009	0.49973585	0.49973588	0.49973584
0.01	0.01	0.49963020	0.49963021	0.49963022

$$u(x,t) = \cfrac{1}{1 + e^{\left(\frac{x}{\sqrt{2}} + \gamma t\right)}}, \tag{9.35}$$

where $\gamma = \frac{1}{\sqrt{2}} - \sqrt{2}\mu$.

Our proposed methods HWM and LLWM can be compared with Wazwaz and Gorguis results (see Ref. [20]), Soliman's results [23], and Mehmet Merdan results (see Ref. [45]).

The numerical solutions of time-fractional Fitzhugh–Nagumo (FN) (Example 9.2) for different values of α, (That is, $\alpha = 0.5$, $\alpha = 0.75$ and $\alpha = 1.0$) and different values of t with $\mu = 0.45$ are presented in Tables 9.1, 9.2, and 9.3. Table 9.4 shows the numerical solutions of time-fractional FN equation for $\alpha = 0.7$, $t = 0.2$, and

Table 9.2 Numerical values when $\alpha = 0.75$ and $\mu = 0.45$ for Example 9.2

x	t	Exact solution u_{HAM}	Numerical u_{HWM} ($m = 32$)	Numerical u_{LLWM} ($k = 1$ and $M = 3$)
0.001	0.001	0.49989967	0.49989969	0.49989966
0.002	0.002	0.49977499	0.49977498	0.49977497
0.003	0.003	0.49964385	0.49964388	0.49964388
0.004	0.004	0.49950898	0.49950889	0.49950892
0.005	0.005	0.49937151	0.49937150	0.49937151
0.006	0.006	0.49923211	0.49923210	0.49923210
0.007	0.007	0.49909115	0.49909111	0.49909113
0.008	0.008	0.49894892	0.49894890	0.49894891
0.009	0.009	0.49880561	0.49880558	0.49880560
0.01	0.01	0.49866137	0.49866132	0.49866132

Table 9.3 Numerical values when $\alpha = 1.0$ and $\mu = 0.45$ for Example 9.2

x	t	Exact solution u_{HAM}	Numerical u_{HWM} ($m = 32$)	Numerical u_{LLWM} ($k = 1$ and $M = 3$)
0.001	0.001	0.49983572	0.49983568	0.49983571
0.002	0.002	0.49967144	0.49967142	0.49967145
0.003	0.003	0.49950716	0.49950715	0.49950715
0.004	0.004	0.49934288	0.49934286	0.49934288
0.005	0.005	0.49917859	0.49917857	0.49917858
0.006	0.006	0.49901431	0.49901431	0.49901430
0.007	0.007	0.49988501	0.49988501	0.49988502
0.008	0.008	0.49868574	0.49868570	0.49868571
0.009	0.009	0.49852145	0.49852142	0.49852146
0.01	0.01	0.49835716	0.49835710	0.49835712

Table 9.4 Numerical values when $\alpha = 0.7$, $t = 0.2$ and $\mu = 0.6$ for Example 9.2

x	Exact solution u_{HAM}	Numerical u_{HWM} ($m = 32$)	Numerical u_{LLWM} ($k = 1$ and $M = 3$)
0.00	0.49065905	0.49065902	0.49065903
0.25	0.44663406	0.44663402	0.44663406
0.50	0.40245145	0.40245144	0.40245144
0.75	0.36175911	0.36175910	0.36175912
1.0	0.32225612	0.32225611	0.32225610

$\mu = 0.6$. The wavelet methods like HWM and LLWM results are in excellent agreement with the exact solution and those obtained by the homotopy analysis method (HAM).

All the numerical experiments presented in this section were computed in double precision with some MATLAB codes on a personal computer System Vostro 1400 Processor x86 Family 6 Model 15 Stepping 13 Genuine Intel ~ 1596 MHz.

9.6 Conclusion

Two reliable wavelet methods have been successfully employed to obtain the numerical solutions of Fitzhugh–Nagumo (FN) and time-fractional FN equations arising in population dynamics. The proposed schemes are the capability to overcome the difficulty arising in calculating the integral values while dealing with nonlinear problems. These two wavelet methods show higher efficiency than the traditional Legendre wavelet method for solving nonlinear PDEs. Numerical example illustrates the power of the proposed schemes. Also, this chapter illustrates the validity and excellent potential of the wavelet methods for nonlinear and fractional PDEs. The numerical solutions obtained using the proposed method show that the solutions are in very good coincidence with the exact solution. In addition, the calculations involved in HWM and LLWM are simple, straightforward, and low computation cost. In Sect. 9.4, we have developed the convergence of the proposed algorithm.

References

1. S.Z. Rida, A.M.A. El-Sayed, A.A.M. Arafa, On the solutions of time-fractional reaction-diffusion equations. Commun. Nonlinear Sci. Numer. Simul. **5**(12), 3847–3854 (2010)
2. A. Cuyt, L. Wuytack, *Nonlinear Methods in Numerical Analysis* (Elsevier Science, Amsterdam, 1987)
3. J.D. Murray, *Lectures on Non-linear Differential Equation Models in Biology* (Clarenden, Oxford, 1977)
4. B.I. Henry, S.L. Wearne, Fractional reaction-diffusion. Physica A **276**(3–4), 448–455 (2000)
5. B. Baeumer, M. Kovács, M.M. Meerschaert, Numerical solutions for fractional reaction-diffusion equations. Comput. Math. Appl. **55**(10), 2212–2226 (2008)
6. S. Momani, R. Qaralleh, Numerical approximations and Padé approximants for a fractional population growth model. Appl. Math. Model. **31**(9), 1907–1914 (2007)
7. G. Hariharan, The homotopy analysis method applied to the Kolmogorov–Petrovskii–Piskunov (KPP) and fractional KPP equations. J. Math. Chem. **51**, 992–1000 (2013). https://doi.org/10.1007/s10910-012-0132-5
8. S.B. Yuste, L. Acedo, K. Lindenberg, Reaction front in an A + B → C reaction-subdiffusion process. Phys. Rev. E **69**(3), part 2, Article ID 036126 (2004)
9. D.J. Aronson, H.F. Weinberg, *Nonlinear Diffusion in Population Genetics Combustion and Never Pulse Propagation* (Springer, New York, 1988)
10. R. Fitzhugh, Impulse and physiological states in theoretical models of nerve membrane. Biophys. J. 1(6), 445–466 (1961)
11. R. Fitzhugh, Biological Engineering, pp 1–85. McGraw-Hill (1969)
12. K. Seki, M. Wojcik, M. Tachiya, Fractional reaction–diffusion equation. J. Chem. Phys. **119**, 2165–2174 (2003)
13. N.A. Khan, N.-U. Khan, A. Ara, M. Jamil, Approximate analytical solutions of fractional reaction-diffusion equations. J. King Saud Univ. Sci. 24, 111–118 (2012)
14. A. Slavova, P. Zecca, CNN model for studying dynamics and traveling wave solutions of FitzHugh-Nagumo equation. J. Comput. Appl. Math. **151**, 13–24 (2003)
15. A.L. Hodgkin, A.F. Huxley, Aquantitive description of membrane current and its application to conduction and excitation in nerve. J. Physiol. **117**, 500 (1952)
16. A. Panfilov, P. Hogeweg, Spiral breakup in a modified Fitzhugh-Nagumo model. Phys. Lett. A **176**, 295–299 (1993)
17. J.S. Nagumo, S. Arimoto, S. Yoshizawa, An active pulse transmission line simulating nerve axon. Proc. IRE **50**, 2061–2071 (1962)
18. H.C. Rosu, O. Cornejo-Perez, Super symmetric pairing of kinks for polynomial nonlinearities. Phys. Rev. E **71**, 1–13 (2005)
19. G. Hariharan, K. Kannan, Haar wavelet method for solving Fisher's equation. Appl. Math. Comput. **211**, 284–292 (2009)
20. A.M. Wazwaz, A. Gorguis, An analytical study of Fisher's equation by using Adomian decomposition method. Appl. Math. Comput. **154**, 609–620 (2004)

21. D. Olmos, B. Shizgal, Pseudospectral method of solution of the Fitzhugh-Nagumo equation. Math. Comput. Simul. **79**, 2258–2278 (2009)
22. D. Olmos, B. Shizgal, A spectral method of solution of Fisher's equation. J. Comput. Appl. Math. **193**, 219–242 (2006)
23. A.A. Soliman, Numerical simulation of the FitzHugh-Nagumo equations. Abstract Appl. Anal. **2012**, Article ID 762516, 13 p. https://doi.org/10.1155/2012/762516
24. S.J. Liao, *Beyond Perturbation: Introduction to Homotopy Analysis Method* (CRC Press/ Chapman and Hall, Boca Raton, 2004)
25. S. Abbasbandy, Soliton solutions for the Fitzhugh-Nagumo equation with the homotopy analysis method. Appl. Math. Model. **32**, 2706–2714 (2008)
26. G. Hariharan, K. Kannan, K. Sharma, Haar wavelet in estimating the depth profile of soil temperature. Appl. Math. Comput. **210**, 119–225 (2009)
27. G. Hariharan, K. Kannan, Haar wavelet method for solving nonlinear parabolic equations. J. Math. Chem. **48**, 1044–1061 (2010)
28. G. Hariharan, K. Kannan, A comparative study of a haar wavelet method and a restrictive Taylor's series method for solving convection-diffusion equations. Int. J. Comput. Methods Eng. Sci. Mech. **11**(4), 173–184 (2010)
29. G. Hariharan, K. Kannan, Haar Wavelet Method for Solving FitzHugh-Nagumo Equation, vol. 43 (World Academy of Science, Engineering and Technology, 2010)
30. U. Lepik, Numerical solution of evolution equations by the Haar wavelet method. J. Appl. Math. Comput. **185**, 695–704 (2007)
31. U. Lepik, Solving PDEs with the aid of two-dimensional Haar wavelets. Comput. Math Appl. **61**, 1873–1879 (2011)
32. H. Jafari, M. Soleymanivaraki, M.A. Firoozjaee, Legendre wavelets for solving fractional differential equations. J. Appl. Math. 7(427), 65–70 (2011)
33. Y. Yang, Solving a nonlinear multi-order fractional differential equation using Legendre pseudo-spectral method. Appl. Math. **4**, 113–118 (2013). https://doi.org/10.4236/am.2013.41020
34. M.H. Heydari, M.R. Hooshmandasl, F.M. Maalek Ghaini, F. Mohammadi, Wavelet collocation method for solving multiorder fractional differential equations. J. Appl. Math. **2012** (2012), Article ID 54240. https://doi.org/10.1155/2012/542401
35. H.A. Abdusalam, Analytic and approximate solutions for Nagumo telegraph reaction diffusion equation. Appl. Math. Comput. **157**, 515–522 (2004)
36. D.Y. Chen, Y. Gu, Cole-Hopf quotient and exact solutions of the generalized Fitzhugh-Nagumo equations. Acta Math. Sci. **19**(1), 7–14 (1999)
37. H. Li, Y. Guo, New exact solutions to the Fitzhugh-Nagumo equation. Appl. Math. Comput. **180**, 524–528 (2006)
38. M. Shih, E. Momoniat, F.M. Mahomed, Approximate conditional symmetries and approximate solutions of the perturbed Fitzhugh-Nagumo equation. J. Math. Phys. **46**, 023503 (2005)
39. M. Razzaghi, S. Yousefi, The Legendre wavelets operational matrix of integration. Int. J. Syst. Sci. **32**, 495–502 (2001)
40. M. Razzaghi, S. Yousefi, The Legendre wavelets direct method for variational problems. Math. Comput. Simulat. **53**, 185–192 (2000)
41. F. Mohammadi, M.M. Hosseini, A new Legendre wavelet operational matrix of derivative and its applications in solving the singular ordinary differential equations. J. Franklin Inst. **348**, 1787–1796 (2011)
42. H. Parsian, Two dimension Legendre wavelets and operational matrices of integration. Acta Math. Acad. Paedagog. Nyhaza. **21**, 101–106 (2005)
43. S.A. Yousefi, Legendre wavelets method for solving differential equations of Lane-Emden type. App. Math. Comput. **181**, 1417–1442 (2006)
44. F. Yin, J. Song, F. Lu, H. Leng, A coupled method of Laplace transform and Legendre wavelets for Lane-Emden-type differential equations. J. Appl. Math. **2012**, Article ID 163821 (2012). https://doi.org/10.1155/2012/163821

45. M. Merdan, Solutions of time-fractional reaction-diffusion equation with modified Riemann-Liouville derivative. Int. J. Phys. Sci. **7**(15), 2317–2326 (2012)
46. V. Turut, N. Guzel, Comparing numerical methods for solving time-fractional reaction-diffusion equations. ISRN Math. Anal. **2012**, Article ID 737206. https://doi.org/10.5402/2012/737206
47. F.C. Meral, T.J. Royston, R. Magin, Fractional calculus in viscoelasticity: an experimental study. Commun. Nonlinear Sci. Numer. Simul. **15**(4), 939–945 (2010)
48. W. Malfliet, Solitary wave solutions of nonlinear wave equations. Am. J. Phys. **60**(7), 650–654 (1992)
49. M.D. Bramson, Maximal displacement of branching Brownian motion. Commun. Pure Appl. Math. **31**(5), 531–581 (1978)
50. K. Maleknejad, S. Sohrabi, Numerical solution of Fredholm integral equations of the first kind by using Legendre wavelets. Appl. Math. Comput. **186**, 836–843 (2007)
51. S.K. Elagan, M. Sayed, Y.S. Hamed, An innovative solutions for the generalized FitzHugh-Nagumo equation by usin the generalized $(G'G)$-expansion method. Appl. Math. **2**, 470–474 (2011)
52. A. Hajipour, S. Molla Mahmoudi, Application of Exp-function method to Fitzhugh-Nagumo equation. World Appl. Sci. J. **9**(1), 113–117 (2010)
53. C.F. Chen, C.H. Hsiao, Haar wavelet method for solving lumped and distributed-parameter systems. IEEE Proc.: Part D **144**(1), 87–94 (1997)

Chapter 10
A New Coupled Wavelet-Based Method Applied to the Nonlinear Reaction–Diffusion Equation Arising in Mathematical Chemistry

In this chapter, we have applied the wavelet-based coupled method for finding the numerical solution of the Murray equation. To the best of our knowledge, until now there is no rigorous Legendre wavelets solution has been reported for the Murray equation. The highest derivative in the differential equation is expanded into Legendre series, and this approximation is integrated while the boundary conditions are applied using integration constants. With the help of Legendre wavelets operational matrices, the Murray equation is converted into an algebraic system. Block-pulse functions are used to investigate the Legendre wavelets coefficient vectors of nonlinear terms. The convergence of the proposed method is proved. Finally, we have given a numerical example to demonstrate the validity and applicability of the method. Moreover, the use of proposed wavelet-based coupled method is found to be simple, efficient, less computation costs, and computationally attractive.

10.1 Introduction

Wavelet analysis, as a relatively new and emerging area in applied mathematical Research, has received considerable attention in dealing with PDEs [1–4]. In recent years, wavelet transforms have found their way into many different fields in science and engineering [3, 4]. Moreover, wavelet transform methods establish a connection with fast numerical algorithms. Analytical methods enable researchers to study the effect of different variables or parameters on the function under study easily. Recently, many new approaches to NLPDEs have been proposed, for example, the Adomian decomposition method [5] and homotopy analysis method [6, 7].

In the numerical analysis, wavelet-based methods and hybrid methods become important tools because of the properties of localization. In wavelet-based methods, there are two important ways of improving the approximation of the solutions: Increasing the order of the wavelet family and the increasing the resolution level of

© Springer Nature Singapore Pte Ltd. 2019

G. Hariharan, *Wavelet Solutions for Reaction–Diffusion Problems in Science and Engineering*, Forum for Interdisciplinary Mathematics, https://doi.org/10.1007/978-981-32-9960-3_10

the wavelet. There is a growing interest in using various wavelets to study problems, of greater computational complexity. Among the wavelet transform families, the Haar and Legendre wavelets deserve much attention. The basic idea of the Legendre wavelet method is to convert the PDEs to a system of algebraic equations by the operational matrices of integral or derivative [8–10]. The main goal is to show how wavelets and multi-resolution analysis can be applied for improving the method in terms of easy implementability and achieving the rapidity of its convergence. Razzaghi and Yousefi [10] introduced the Legendre wavelet method for solving variational problems and constrained optimal control problems. Yousefi [11] established the Legendre wavelet method for solving the differential equations of Lane–Emden type. Hariharan et al. [1–4, 12] had introduced the Haar wavelet method for diffusion equation, convection—diffusion equation, reaction–diffusion equation, nonlinear parabolic equations, fractional Klein–Gordon equations, sine–Gordon equations, KPP equation and Fisher type equations. Mohammadi and Hosseini [13] had showed a new Legendre wavelet operational matrix of derivative in solving singular ordinary differential equations. Parsian [9] introduced two-dimensional Legendre wavelets and operational matrices of integration.

In recent years, nonlinear reaction–diffusion equations (NLRDEs) have been widely studied and applied in biological science and engineering [14–17]. This study concerns the numerical solutions of nonlinear reaction–diffusion modeling the dynamics of diffusion and nonlinear reproduction for a population [18–22]. The associated nonlinear reaction–diffusion equation was initiated by Fisher [16] to describe the propagation behavior of a virile mutant. The nonlinear reaction–diffusion equations describe a population of diploid individuals [14, 16].

In this work, we have applied a wavelet-based coupled method (LLWM) which combines the Laplace transform method and the Legendre wavelets method for the numerical solution of Murray equation.

10.2   Legendre Wavelets and Properties

10.2.1   Wavelets

Wavelets are the family of functions which are derived from the family of scaling function $\{\emptyset_{j,k}: k \in Z\}$ where:

$$\emptyset(x) = \sum_k a_k \emptyset(2x - k) \tag{10.1}$$

For the continuous wavelets, the following equation can be represented:

$$\Psi_{a,b}(x) = |a|^{-\frac{1}{2}}\Psi\left(\frac{x-b}{a}\right) \, a, b \in R, \, a \neq 0. \tag{10.2}$$

where a and b are dilation and translation parameters, respectively, such that $\Psi(x)$ is a single wavelet function.

The discrete values are put for a and b in the initial form of the continuous wavelets, i.e.,

$$a = a_0^{-j}, a_0 > 1, b_0 > 1, \tag{10.3}$$

$$b = kb_0 a_0^{-j}, \quad j, k \in Z. \tag{10.4}$$

Then, a family of discrete wavelets can be constructed as follows:

$$\Psi_{j,k} = |a_0|^{\frac{1}{2}}\Psi(2^j x - k), \tag{10.5}$$

So, $\Psi_{j,k}(x)$ constitutes an orthonormal basis in $L^2 (R)$, where $\Psi(x)$ is a single function.

10.2.2   Legendre Wavelets

The Legendre wavelets are defined by

$$\psi_{nm}(t) = \begin{cases} \sqrt{m + \frac{1}{2}}2^{\frac{k}{2}}L_m\left(2^k t - \hat{n}\right), & \text{for} \quad \frac{\hat{n}-1}{2^k} \leq t \leq \frac{\hat{n}+1}{2^k}, \\ 0, & \text{otherwise} \end{cases} \tag{10.6}$$

where $m = 0, 1, 2, ..., M - 1$ and $k = 1, 2, ..., 2^{j-1}$. The coefficient $\sqrt{m + \frac{1}{2}}$ is for orthonormality, and then, the wavelets $\Psi_{k,m}(x)$ form an orthonormal basis for L^2 [0, 1]. In the above formulation of the Legendre wavelets, the Legendre polynomials are in the following way:

$$\begin{aligned} p_0 &= 1, \\ p_1 &= x, \\ p_{m+1}(x) &= \frac{2m+1}{m+1}x p_m(x) - \frac{m}{m+1}p_{m-1}(x). \end{aligned} \tag{10.7}$$

and $\{p_{m+1}(x)\}$ are the orthogonal functions of order m, which is named the well-known shifted Legendre polynomials on the interval [0, 1]. Note that, in the general form of the Legendre wavelets, the dilation parameter is $a = 2^{-j}$ and the translation parameter is $b = n \, 2^j$ [8].

10.2.3   Two-dimensional Legendre Wavelets

Two-dimensional Legendre wavelets in $L^2(R)$ over the interval $[0, 1] \times [0, 1]$ as the form

$$\Psi_{n,m,n',m'}(x, y) = \begin{cases} \sqrt{\left(m + \frac{1}{2}\right)\left(m' + \frac{1}{2}\right)} 2^{\frac{k+k'}{2}} p_m(x) p_{m'}(y), \\ \quad\quad \frac{n-1}{2^{k-1}} \leq x \leq \frac{n}{2^{k-1}}, \frac{n'-1}{2^{k-1}} \leq y \leq \frac{n'}{2^{k'-1}}; \\ 0, \quad\quad\quad\quad\quad\quad \text{otherwise.} \end{cases} \quad (10.8)$$

and $m = 0, 1, 2, ..., M - 1, m' = 0, 1, 2, 3, ... M' - 1, n = 1, 2, ..., 2^{k-1}, n' = 1, 2, ...2^{k'-1}$

$$\text{where } P_m(x) = \overline{P_{m'}}\left(2^k x - 2n + 1\right), P_{m'}(y) = \overline{P_{m'}}\left(2^k y - 2n' + 1\right), \quad (10.9)$$

$\overline{P_m}$ are the Legendre functions of order m defined over the interval $[-1, 1]$.

By using the two-dimensional shifted Legendre polynomials into $x \in \left[\frac{n-1}{2^{k-1}}, \frac{n}{2^{k-1}}\right]$ and

$y \in \left[\frac{n'-1}{2^{k'-1}}, \frac{n'}{2^{k'-1}}\right]$, the $\int_0^1 \Psi_{n,m,n',m'}(x, y)$ can be written as

$$\int_0^1 \Psi_{n,m,n',m'}(x, y) = A_{m,m'}.P_{m'}(x) P_{m'}(y) \chi_{\left[\begin{smallmatrix} \frac{n-1}{2^{k-1}}, \frac{n}{2^{k-1}} \\ \frac{n'-1}{2^{k'-1}}, \frac{n'}{2^{k'-1}} \end{smallmatrix}\right]}(x, y), \quad (10.10)$$

in which $A_{m,m'} = \sqrt{\left(m + \frac{1}{2}\right)\left(m' + \frac{1}{2}\right)} 2^{\frac{k+k'}{2}}$ and $\chi_{\left[\begin{smallmatrix} \frac{n-1}{2^{k-1}}, \frac{n}{2^{k-1}} \\ \frac{n'-1}{2^{k'-1}}, \frac{n'}{2^{k'-1}} \end{smallmatrix}\right]}(x, y)$ are a character-

istic function defined as $\chi_{\left[\begin{smallmatrix} \frac{n-1}{2^{k-1}}, \frac{n}{2^{k-1}} \\ \frac{n'-1}{2^{k'-1}}, \frac{n'}{2^{k'-1}} \end{smallmatrix}\right]}(x, y) = \begin{cases} 1, & x \in \left[\frac{n-1}{2^{k-1}}, \frac{n}{2^{k-1}}\right], y \in \left[\frac{n'-1}{2^{k'-1}}, \frac{n'}{2^{k'-1}}\right]; \\ 0, & \text{otherwise} \end{cases}$

The two-dimensional Legendre wavelets are an orthonormal set over $[0, 1] \times [0, 1]$.

$$\int_0^1 \int_0^1 \Psi_{n,m,n',m'}(x, y) \Psi_{n_1,m_1,n'_1,m'_1}(x, y) dx dy = \delta_{n,n_1} \delta_{n',n'_1} \delta_{m',m'_1} \quad (10.11)$$

The function $u(x, y) \in L^2(R)$ defined over $[0, 1] \times [0, 1]$ may be expanded as

$$u(x, y) = X(x)Y(y) \cong \sum_{n=1}^{\infty}\sum_{m=0}^{\infty}\sum_{n'=1}^{\infty}\sum_{m'=0}^{\infty} c_{n,m,n',m'}\Psi_{n,m,n',m'}(x, y) \tag{10.12}$$

If the infinite series in Eq. (10.12) is truncated, then Eq. (10.13) can be written as

$$u(x, y) = X(x)Y(y) \cong \sum_{n=1}^{2^{k-1}}\sum_{m=0}^{M-1}\sum_{n'=1}^{2^{k'-1}}\sum_{m'=0}^{M'-1} c_{n,m,n',m'}\Psi_{n,m,n',m'}(x, y) \tag{10.13}$$

where $c_{n,m,n',m'} = \int_0^1\int_0^1 X(x)Y(y)\Psi_{n,m,n',m'}(x, y)\mathrm{d}x\mathrm{d}y$.

Equation (10.13) can be expressed as the form

$$u(x, y) = c^{\mathrm{T}}.\Psi(x, y) \tag{10.14}$$

where C and $\Psi(x, y)$ are coefficients matrix and wavelets vector matrix, respectively. The number of dimensions of C and $\Psi(x, y)$ are $2^{k-1}2^{k'-1}MM' \times 1$, and given by

$$
\begin{aligned}
C = \Big[& c_{1,0,1,0}, \cdots c_{1,0,1,M'-1}, c_{1,0,2,0}, \cdots, c_{1,0,2,M'-1}, \cdots, c_{1,0,2^{k'-1},0}, \cdots, \\
& c_{1,0,2^{k'-1},M'-1}, \cdots c_{1,M-1,1,0}, \cdots c_{1,M-1,1,M'-1}, c_{1,M-1,2,0}, \cdots, \\
& c_{1,M-1,2,M'-1}, \cdots, c_{1,M-1,2^{K-1},0}, \cdots c_{1,M-1,2^{K-1},M'-1}, \cdots c_{2,0,1,0}, \cdots \\
& c_{2,0,1,M'-1}, c_{2,0,2,0}, \cdots, c_{2,0,2,M'-1}, \cdots c_{2,0,2^{k-1},0}, \cdots c_{2,0,2^{k-1},M'-1}, \cdots \\
& c_{2,M-1,1,0} \cdots c_{2,M-1,1,M'-1}, c_{2,M-1,2,0}, \cdots, c_{2,M-1,2,M'-1}, \cdots, \\
& c_{2,M-1,2^{k-1},0}, \cdots, c_{2,M-1,2^{k-1},M'-1}, \cdots, c_{2^{k-1},0,1,0}, \cdots c_{2^{k-1},0,1M'-1}, \\
& c_{2^{k-1},0,2,0}, \cdots, c_{2^{k-1},0,,M'-1}, \cdots c_{2^{k-1},0,2^{k-1},0}, \cdots c_{2^{k-1},M-1,2^{k'-1},M'-1} \Big]^{\mathrm{T}}
\end{aligned} \tag{10.15}
$$

$$
\begin{aligned}
\Psi = \Big[& \Psi_{1,0,1,0}, \cdots, \Psi_{1,0,1,M'-1}, \Psi_{1,0,2,0}, \cdots \Psi_{1,0,2^{k-1},0}, \cdots \\
& \Psi_{1,0,2^{k'-1},M'-1}, \cdots, \Psi_{1,M-1,1,0}, \cdots \Psi_{1,M-1,1,M'-1}, \\
& \Psi_{1,M-1,2,0}, \cdots, \Psi_{1,M-1,2,M'-1}, \cdots \Psi_{1,M-1,2^{k-1},0}, \cdots, \\
& \Psi_{1,M-1,2^{k-1},M'-1}, \cdots, \Psi_{2,0,1,0}, \cdots, \Psi_{2,0,1,M'-1}, \\
& \Psi_{2,0,2,0}, \cdots \Psi_{2,0,2,M'-1}, \cdots, \Psi_{2,0,2^{k'-1},0}, \cdots, \Psi_{2,0,2^{k-1},M'-1}, \cdots, \\
& \Psi_{2,M-1,1,0}, \cdots, \Psi_{2,M-1,1,M'-1}, \Psi_{2,M-1,2,0}, \cdots, \\
& \Psi_{2,M-1,2,M'-1}, \cdots, \Psi_{2,M-1,2^{k'-1},0}, \cdots, \Psi_{2,M-1,2^{k'-1},M'-1}, \\
& \Psi_{2^{k-1},0,1,0}, \cdots, \Psi_{2^{k-1},0,1,M'-1}, \Psi_{2^{k-1},0,2,0}, \cdots, \\
& \Psi_{2^{k-1},0,2,M'-1}, \cdots, \Psi_{2^{k-1},0,2^{k-1},0}, \cdots \Psi_{2^{k-1},M-1,2^{k-1},M'-1} \Big]^{\mathrm{T}}
\end{aligned} \tag{10.16}
$$

The integration of the product of two Legendre wavelet function vectors is obtained as

$$\int_0^1 \int_0^1 \Psi(x,y)\Psi^T(x,y)\mathrm{d}x\mathrm{d}y = I \tag{10.17}$$

where I is the identity matrix.

Another form of the two-dimensional Legendre wavelets by using the one-dimensional Legendre wavelets was given in [8].

A two-dimensional function $f(x, y)$ defined $[0, 1) \times [0, 1)$ may be expanded by Legendre wavelet series as

$$f(x,y) = \sum_{i=1}^{2^k M} \sum_{j=1}^{2^k M} C_{ij}\Psi_i(x)\Psi_j(y) = \Psi^T(x)C\,\Psi(y), \tag{10.18}$$

where

$$C_{ij} = \int_0^1 f(x,y)\Psi_i(x)\mathrm{d}x \int_0^1 f(x,y)\Psi_j(y)\mathrm{d}t \tag{10.19}$$

Equation (10.18) can be written into the discrete form (in matrix form) by

$$f(x,y) = \Psi^T(x)C\Psi(y), \tag{10.20}$$

where C and $\Psi(t)$ are $2^{k-1}M \times 1$ matrices given by

$$C = \begin{bmatrix} c_{0,0} & c_{0,1} & \cdots & c_{0,2^{k-1}M} \\ c_{1,0} & c_{1,1} & \cdots & c_{1,2^{k-1}M} \\ \vdots & \vdots & \ddots & \vdots \\ c_{2^{k-1}M,0} & c_{2^{k-1}M,1} & \cdots & c_{2^{k-1}M2^{k-1}M} \end{bmatrix}$$

The two-dimensional Legendre wavelet operational matrix of integration has been derived (See Ref. [8]).

Theorem 10.1 Let $\Psi(x,y)$ be the two-dimensional Legendre wavelets vector.

$$\frac{\partial \Psi(x,y)}{\partial x} = D_x \Psi(x,y), \tag{10.21}$$

where D_x is $2^{k-1,}2^{k'-1}MM' \times 2^{k-1}2^{k'-1}MM'$ and has the form as follows:

$$D_x = \begin{bmatrix} D & O' & \cdots & 0' \\ 0' & D & \cdots & 0' \\ \vdots & \vdots & \ddots & \vdots \\ O' & O' & \cdots & D \end{bmatrix}$$

In which 0' and D are $2^{k-1} 2^{k'-1} MM' \times 2^{k-1} 2^{k'-1} MM'$ matrix and the element of D is defined as follows:

$$D_{r,s} = \begin{cases} 2^k \sqrt{(2r-1)(2s-1)}I, & r = 2,3,\ldots M; s = 1,\ldots r-1; r+s \text{ is odd} \\ 0 & \text{otherwise} \end{cases}$$

$$(10.22)$$

and I and O are $2^{k'-1}M' \times 2^{k'-1}M'$ identity matrices.

Theorem 10.2 Let $\Psi(x,y)$ be the two-dimensional Legendre wavelets vector. Then,

$$\frac{\partial \Psi(x,y)}{\partial x} = D_y \Psi(x,y), \tag{10.23}$$

$$D_y = \begin{bmatrix} D & O' & \cdots & 0' \\ 0' & D & \cdots & 0' \\ \vdots & \vdots & \ddots & \vdots \\ O' & O' & \cdots & D \end{bmatrix},$$

where D_y is $2^{k-1,} 2^{k'-1}MM' \times 2^{k-1}2^{k'-1}MM'$ and O', D is $MM' \times MM'$ matrix is given as

$$D = \begin{bmatrix} F & O & \cdots & 0 \\ 0 & F & \cdots & 0 \\ \vdots & \vdots & \ddots & \vdots \\ O & O & \cdots & F \end{bmatrix},$$

in which O and F are $M' \times M'$ matrix, and F is defined as follows:

$$F_{r,s} = \begin{cases} 2^{k'} \sqrt{(2r-1)(2s-1)}, & r = 2,\ldots, M'; S = 1,\ldots, r-1; \text{ and } r+s \text{ is odd} \\ 0, & \text{otherwise} \end{cases}$$

$$(10.24)$$

By using Eqs. (10.21) and (10.23), the operational matrices for nth derivative can be derived as

$$\frac{\partial^n \Psi(x,y)}{\partial x^n} = D_x^n \Psi(x,y), \frac{\partial^m \Psi(x,y)}{\partial y^m} = D_y^m \Psi(x,y)$$

$$\frac{\partial^{n+m} \Psi(x,y)}{\partial x^n \partial y^m} = D_x^n D_y^m \Psi(x,y),$$

where D^n is the nth power of matrix D.

10.2.4   Block-Pulse Functions (BPFs)

The block-pulse functions form a complete set of orthogonal functions which are defined on the interval $[0, b)$ by

$$b_i(t) = \begin{cases} 1, & \frac{i-1}{m}b \leq t < \frac{i}{m}b, \\ 0, & \text{elsewhere} \end{cases} \tag{10.25}$$

for $i = 1, 2, \ldots, m$. It is also known that for any absolutely integrable function $f(t)$ on $[0, b)$ can be expanded in block-pulse functions:

$$f(t) \cong \xi^{\mathrm{T}} B_m(t) \tag{10.26}$$

$$\xi^{\mathrm{T}} = [f_1, f_2, \ldots, f_m], B_m(t) = [b_1(t), b_2(t), \ldots, b_m(t)], \tag{10.27}$$

where f_i are the coefficients of the block-pulse function, given by

$$f_i = \frac{m}{b} \int_0^b f(t) b_i(t) \mathrm{d}t \tag{10.28}$$

Remark 1: Let A and B are two matrices of $m \times m$, then $A \otimes B = \left(a_{ij} \times b_{ij}\right)_{mm}$.
Lemma 1: Assuming $f(t)$ and $g(t)$ are two absolutely integrable functions, which can be expanded in block-pulse function as $f(t) = FB(t)$ and $g(t) = GB(t)$, respectively, then we have

$$f(t)g(t) = FB(t)B^{\mathrm{T}}(t)G^{\mathrm{T}} = HB(t), \tag{10.29}$$

where $H = F \otimes G$.

10.3   Approximating the Nonlinear Term

The Legendre wavelets can be expanded into m-set of block-pulse functions as

$$\Psi(t) = \emptyset_{m \times m} B_m(t) \tag{10.30}$$

Taking the collocation points as following

$$t_i = \frac{i - 1/2}{2^{k-1}M}, i = 1, 2, \ldots, 2^{k-1}M \tag{10.31}$$

The m-square Legendre matrix $\emptyset_{mm}$ is defined as

$$\emptyset_{mm} \cong [\Psi(t_1)\Psi(t_2)\ldots\Psi(t_{2^{k-1}M})] \tag{10.32}$$

The operational matrix of product of Legendre wavelets can be obtained by using the properties of BPFs, and let $f(x,t)$ and $g(x,t)$ are two absolutely integrable functions, which can be expanded by Legendre wavelets as $f(x,t) = \Psi^T(x)F\Psi(t)$ and $g(x,t) = \Psi^T(x)G\Psi(t)$ respectively.

From Eq. (10.30), we have

$$f(x,t) = \Psi^T(x)F\Psi(t) = B^T(x)\emptyset_{mm}^T F\emptyset_{mm}B(t), \tag{10.33}$$

$$g(x,t) = \Psi^T(x)G\Psi(t) = B^T(x)\emptyset_{mm}^T G\emptyset_{mm}B(t), \tag{10.34}$$

and $F_b = \emptyset_{mm}^T F\emptyset_{mm}, G_b = \emptyset_{mm}^T G\emptyset_{mm}, H_b = F_b \otimes G_b$.

Then,

$$\begin{aligned}
f(x,t)g(x,t) &= B^T H_b B(t),\\
&= B^T(x)\emptyset_{mm}^T \mathrm{inv}\left(\emptyset_{mm}^T\right)H_b\mathrm{inv}\left(\mathrm{inv}\left(\emptyset_{mm}^T\right)H_b\mathrm{inv}(\emptyset_{mm})\right)\emptyset_{mm}B(t)\\
&= \Psi^T(x)H\,\Psi(t)
\end{aligned} \tag{10.35}$$

where $H = \mathrm{inv}\left(\emptyset_{mm}^T\right)H_b\mathrm{inv}((\emptyset_{mm}))$.

10.4 Function Approximation

A given function $f(x)$ with the domain $[0, 1]$ can be approximated by:

$$f(x) = \sum_{k=1}^{\infty}\sum_{m=0}^{\infty} c_{k,m}\Psi_{k,m}(x) = C^T \cdot \Psi(x). \tag{10.36}$$

Here, C and Ψ are the matrices of size $(2^{j-1}M \times 1)$.

$$C = \left[c_{1,0}, c_{1,1}, \ldots c_{1,M-1}, c_{2,0}, c_{2,1}, \ldots c_{2,M-1}, \ldots c_{2^{j-1},1}, \ldots c_{2^{j-1},M-1}\right]^T \tag{10.37}$$

$$\Psi(x) = \left[\Psi_{1,0}, \Psi_{1,1}, \Psi_{2,0}, \Psi_{2,1}, \ldots\Psi_{2,M-1}, \ldots\Psi_{2^{j-1},M-1}\right]^T. \tag{10.38}$$

10.5 Mathematical Model and the Method of Solution

Consider the nonlinear reaction–diffusion equations with convection term of the form [19]

$$\frac{\partial U}{\partial t} = A(U)\frac{\partial^2 U}{\partial x^2} + B(U)\frac{\partial U}{\partial x} + C(U), \quad 0 \le x < 1, \; 0 \le t < 1. \tag{10.39}$$

where $U(x,t)$ is an unknown function, $A(U), B(U)$ and $C(U)$ are arbitrary smooth functions. Equation (10.39) is a well-known nonlinear second-order evolution equation describing various models in biology [19].

When $A(U) = 1$, $B(U) = \mu_1 U$ and $C(U) = \mu_2 U - \mu_3 U^2$
Here, $\mu_1, \mu_2, \mu_3 \in R$.
Equation (10.39) becomes

$$\frac{\partial U}{\partial t} = \frac{\partial^2 U}{\partial x^2} + \mu_1 U\frac{\partial U}{\partial x} + \mu_2 U - \mu_3 U^2, \quad 0 \le x < 1, \; 0 \le t < 1, \tag{10.40}$$

which is called the nonlinear Murray equation with initial condition:

$$U(x,0) = f(x), \quad 0 \le x < 1 \tag{10.41}$$

and mixed boundary conditions

$$\left. \begin{array}{l} U(0,t) = G(t), \quad 0 \le t < 1 \\ \frac{\partial U}{\partial x} = I(t), \; 0 \le t < 1 \end{array} \right\} \tag{10.42}$$

The exact solution for Eq. (10.39) is given by

$$U(x,t) = \frac{\mu_2 + c_1 e^{\left(\gamma^2 t + \gamma x\right)}}{\mu_3 + c_0 e^{-(\mu_2 t)}},$$

where $\gamma = \frac{\mu_3}{\mu_1}$ and $\mu_1 \ne 0$

c_0 is a constant such that $\mu_3 + c_0 e^{\left(-\lambda^2 t\right)} \ne 0$ and c_1 is an arbitrary constant. Taking the Laplace transform on both sides, we get

$$sL(U) - U(x,0) = L\left[U_{xx} + \mu_1 UU_x + \mu_2 U - \mu_3 U^2\right] \tag{10.43}$$

$$sL(U) = U(x,0) + L\left[U_{xx} + \mu_1 UU_x + \mu_2 U - \mu_3 U^2\right] \tag{10.44}$$

$$L(U) = \frac{U(x,0)}{s} + \frac{1}{s}L\left[U_{xx} + \mu_1 UU_x + \mu_2 U - \mu_3 U^2\right] \tag{10.45}$$

Taking the inverse Laplace transform to Eq. (10.45), we get

$$U(x,t) = U(x,0) + L^{-1}\left(\frac{1}{s}L\left[U_{xx} + \mu_1 UU_x + \mu_2 U - \mu_3 U^2\right]\right) \tag{10.46}$$

Because

$$L^{-1}\left[\frac{1}{s}L(t^n)\right] = L^{-1}\left(\frac{n!}{s^{n+2}}\right)$$

$$= \frac{1}{n+1}t^{n+1}; (n = 0, 1, 2, \ldots) \tag{10.47}$$

we have

$$L^{-1}\left[s^{-1}L(.)\right] = \int_0^t (.)dt \tag{10.48}$$

From Eq. (10.46),

$$U(x,t) = U(x,0) + L^{-1}\left(\frac{1}{s}L(U_{xx} + g(U))\right) \tag{10.49}$$

where $g(U) = \mu_1 UU_x + \mu_2 U - \mu_3 U^2$ By using the Legendre wavelets method,

$$\left.\begin{array}{l} U(x,t) = C^{\mathrm{T}}\psi(x,t) \\ U(x,0) = S^{\mathrm{T}}\psi(x,t) \\ g(U) = G^{\mathrm{T}}\psi(x,t) \end{array}\right\} \tag{10.50}$$

Substituting Eq. (10.50) in Eq. (10.49), we obtain

$$C^{\mathrm{T}} = S^{\mathrm{T}} + \left(C^{\mathrm{T}}Dx^2 + G^{\mathrm{T}}\right)P_t^2 \tag{10.51}$$

Here, G^{T} has a nonlinear relation with C. When we solve a nonlinear algebraic system, we get the solution is more complex and large computation time. In order to overcome the above drawbacks, we introduce an approximation formula as follows:

$$U_{n+1} = U(x,0) + \Pi\left[\frac{\partial^2 U_n}{\partial x^2} + g(U_n)\right], \tag{10.52}$$

where $g(U) = \mu_1 UU_x + \mu_2 U - \mu_3 U^2$

Expanding $u(x, t)$ by the Legendre wavelets using the following relation

$$C_{n+1}^{\mathrm{T}} = C_0^{\mathrm{T}} + \left[C_n^{\mathrm{T}}D_x^2 + G_n^{\mathrm{T}}\right]P_t^2 \tag{10.53}$$

10.6   Convergence Analysis

$$U^* = U_0 + \Pi\left[U^*_{xx} + g(U^*)\right] \tag{10.54}$$

$$U_{n+1} = U_0 + \Pi\left[(U_n)_{xx} + g(U_n)\right] \tag{10.55}$$

Subtracting Eq. (10.55) from Eq. (10.54), we obtain

$$U_{n+1} - U^* = \Pi\left[(U_n - U^*)_{xx} + g(U_n) - g(U^*)\right] \tag{10.56}$$

Using the Lipschitz condition,

$\|g(U_n) - g(U^*)\| \leq \gamma \|U_n - U^*\|$, we have

$$\|U_{n+1} - U^*\| \leq \left\|\Pi(U_n - U^*)_{xx}\right\| + \left\|\Pi(g(U_n) - g(U^*))\right\| \tag{10.57}$$

$$\leq \left\|\Pi(U_n - U^*)_{xx}\right\| + \gamma\left\|\Pi(U_n - U^*)\right\| \tag{10.58}$$

Let $U_{n+1} = C^T_{n+1}\psi(x,t)$

$$\begin{aligned}
U^* &= C^T\psi(x,t) \\
\in^T_{n+1} &= C^T_{n+1} - C^T \\
\in^T_{n+1} &\leq \in^T_n \left\|D^2_x P^2_t + \gamma P^2_t\right\|
\end{aligned} \tag{10.59}$$

The following formula Eq. (10.59) can be obtained by using recursive relation.

$$\in^T_{n+1} \leq \in^T_n \left\|D^2_x P^2_t + \gamma P^2_t\right\|^n \in_0 \tag{10.60}$$

When $\mathrm{Lim}_{n\to\infty}\left\|D^2_x P^2_t + \gamma P^2_t\right\| = 0^n$, the series solution of Eq. (10.60) using the LLWM converges to $u^*(x)$. By using the definitions of D_x and P_t, we can get the value of γ.

Suppose $k = k' = 1$ and $M = M'$, the maximum element of D_x and P_t is

$2\sqrt{(2M-1)(2M-3)}$ and 0.5, respectively.

10.7   Illustrative Example

Example 10.1 Consider the Murray equation [20]

$$\frac{\partial U}{\partial t} = \frac{\partial^2 U}{\partial x^2} + U\frac{\partial U}{\partial x} + U - U^2, \quad 0 \leq x < 1,\ 0 \leq t < 1 \tag{10.61}$$

With the initial condition

$$U(x,0) = \frac{\mu_2 + c_1 e^{(\gamma x)}}{\mu_3 + c_0} \tag{10.62}$$

and mixed boundary conditions

$$U(0,t) = \frac{\mu_2 + c_1 e^{(\gamma^2 t)}}{\mu_3 + c_0 e^{(-\mu_2 t)}} \tag{10.63}$$

$$\frac{\partial U(0,t)}{\partial x} = \frac{c_1 \gamma e^{(\gamma^2 t)}}{\mu_3 + c_0 e^{(-\mu_2 t)}} \tag{10.64}$$

with $c_0 = 1, c_1 = 1$ and $\gamma = 1$.

We start the first iteration; an initial guess of the solution of u_0 is required. We select $u_0 = u(x,0)$, and expanding u by the Legendre wavelets, we gain

The Legendre wavelets scheme is given by

$$C_{n+1}^T = C_0^T + \left[C_n^T D_x^2 + G_n^T \right] P_t^2$$

Our proposed method (LLWM) can be compared with Cherniha's results [See Ref. [20]]. Good agreement with the exact solution is observed.

More efficient and accurate results can be obtained by using larger values of M. Comparison with these algorithms shows that the LLWM is competitive and efficient.

The numerical solutions of Murray's equation (Example. 10.1) for different values of x and t are presented in Table 10.1. Our LLWM results are in excellent agreement with the exact solution and those obtained by the Cherniha's method [20].

All the numerical experiments presented in this section were computed in double precision with some MATLAB codes on a personal computer System Vostro 1400 Processor x86 Family 6 Model 15 Stepping 13 Genuine Intel $\sim$1596 MHz.

Table 10.1 A comparison between the exact and the LLWM for various values of (x, t) and $M = 4$

x	t	Exact solution $U(x, t)$	Numerical solution u_{LLWM}
0.125	0.125	1.21329571	1.21329569
0.125	0.875	2.62430764	2.62430765
0.375	0.125	1.40702557	1.40702556
0.375	0.875	3.16921683	3.16921680
0.625	0.125	1.65577963	1.65577962
0.625	0.875	3.86889407	3.86889405
0.875	0.125	1.97518616	1.97518615
0.875	0.875	4.76729744	4.76729740

10.8   Conclusion

In this work, a new coupled wavelet-based method has been successfully employed to obtain the numerical solutions of Murray's equation arising in mathematical biology. The proposed scheme is the capability to overcome the difficulty arising in calculating the integral values while dealing with nonlinear problems. This method shows higher efficiency than the traditional Legendre wavelet method for solving nonlinear PDEs. Numerical example illustrates the power of the proposed scheme LLWM. Also, this chapter illustrates the validity and excellent potential of the LLWM for nonlinear PDEs. The numerical solutions obtained using the proposed method show that the solutions are in very good coincidence with the exact solution. In addition, the calculations involved in LLWM are simple, straight forward, and low computational cost. In Sect. 10.4, we have developed the convergence of the proposed algorithm.

References

1. G. Hariharan, K. Kannan, K.R. Sharma, Haar wavelet in estimating the depth profile of soil temperature. Appl. Math. Comput. **210**, 119–225 (2009). https://doi.org/10.1016/j.amc.2008.12.036
2. G. Hariharan, K. Kannan, Haar wavelet method for solving Fisher's equation. Appl. Math. Comput. **211**, 284–292 (2009). https://doi.org/10.1016/j.amc.2008.12.089
3. G. Hariharan, K. Kannan, Haar wavelet method for solving nonlinear parabolic equations. J. Math. Chem. **48**, 1044–1061 (2010). https://doi.org/10.1007/s10910-010-9724-0
4. G. Hariharan, K. Kannan, A comparative study of a Haar wavelet method and a restrictive Taylor's series method for solving convection-diffusion equations. Int. J. Comput. Methods Eng. Sci. Mech. **11**(4), 173–184 (2010). https://doi.org/10.1080/15502281003762181
5. A.M. Wazwaz, A. Gorguis, An analytical study of Fisher's equation by using Adomian decomposition method. Appl. Math. Comput. **154**, 609–620 (2004). https://doi.org/10.1016/S0096-3003(03)00738-0
6. S.J. Liao, *Beyond Perturbation: Introduction to Homotopy Analysis Method* (CRC Press/Chapman and Hall, Boca Raton, 2004)
7. S. Abbasbandy, Soliton solutions for the Fitzhugh-Nagumo equation with the homotopy analysis method. Appl. Math. Model. **32**, 2706–2714 (2008). https://doi.org/10.1016/j.apm.2007.09.019
8. M. Razzaghi, S. Yousefi, The Legendre wavelets operational matrix of integration. Int. J. Syst. Sci. **32**, 495–502 (2001). https://doi.org/10.1080/00207720120227
9. H. Parsian, Two dimension Legendre wavelets and operational matrices of integration. Acta Math. Academiae Paedagogicae Nyiregyháziens **21**, 101–106 (2005)
10. M. Razzaghi, S. Yousefi, The Legendre wavelets direct method for variational problems. Math. Comput. Simulat. **53**, 185–192 (2000). https://doi.org/10.1016/s0378-4754(00)00170-1
11. S.A. Yousefi, Legendre wavelets method for solving differential equations of Lane-Emden type. App. Math. Comput. **181**, 1417–1442 (2006). https://doi.org/10.1016/j.amc.2006.02.031
12. G. Hariharan, The homotopy analysis method applied to the Kolmogorov–Petrovskii–Piskunov (KPP) and fractional KPP equations. J. Math. Chem. **51**, 992–1000 (2013). https://doi.org/10.1007/s10910-012-0132-5

13. F. Mohammadi, M.M. Hosseini, A new Legendre wavelet operational matrix of derivative and its applications in solving the singular ordinary differential equations. J. Franklin Inst. **348**, 1787–1796 (2011). https://doi.org/10.1016/j.jfranklin.2011.04.017

14. J.D. Murray, *Nonlinear Differential Equation Models In Biology* (Clarendon Press, Oxford, 1977)

15. J.D. Murray, *Mathematical Biology* (Springer, Berlin, 1989)

16. R.A. Fisher, The wave of advance of advantageous genes. Ann. Eugenics **7**, 353–369 (1937)

17. R. Cherniha, Symmetry and exact solutions of heat-and-mass transfer equations in Tokamak plasma. Dopovidi Akad. Nauk Ukr. **4**, 17–21 (1995). https://doi.org/10.1007/bf02595363

18. R. Cherniha, A constructive method for construction of new exact solutions of nonlinear evolution equations. Rep. Math. Phys. **38**, 301–310 (1996). https://doi.org/10.1007/bf02513431

19. R. Cherniha, Application of a constructive method for construction of non-lie solutions of nonlinear evolution equations, Ukr. Math. J. **49**: 814–827. https://doi.org/10.1007/bf02513431

20. R.M. Cherniha, New ansatze and exact solution for nonlinear reaction-diffusion equations arising in mathematical biology. Symmetry Nonlinear Math. Phys. **1**, 138–146 (1997)

21. D.J. Aronson, H.F. Weinberg, *Nonlinear Diffusion in Population Genetics Combustion and Never Pulse Propagation* (Springer, New York, NY, USA, 1988)

22. A. Cuyt, L. Wuytack, *Nonlinear Methods in Numerical Analysis* (Elsevier Science, Amsterdam, The Netherland, 1987)

Chapter 11
Wavelet-Based Analytical Expressions to Steady-State Biofilm Model Arising in Biochemical Engineering

In this chapter, we have developed an efficient wavelet-based approximation method to biofilm model under steady state arising in enzyme kinetics. Chebyshev wavelet-based approximation method is successfully introduced in solving non-linear steady-state biofilm reaction model. Analytical solutions for substrate concentration have been derived for all values of the parameters δ and S_L. The power of the manageable method is confirmed. Some numerical examples are presented to demonstrate the validity and applicability of the wavelet method. Moreover, the use of Chebyshev wavelets is found to be simple, efficient, flexible, convenient, small computation costs, and computationally attractive.

11.1 Introduction

In recent years, biofilms play a very important role in many scientific and technological areas. A biofilm is a well-organized, cooperating community of microorganisms attached to an environmental surface. These surfaces include biological tissues, water surfaces, and solid substrates which can be located in marine or freshwater environments. They play significant roles in many scientific and engineering fields. The diffusion-controlled kinetics of substrate conversion was analyzed under the assumption that the reaction proceeds via a simple mechanism and obeys Michealis–Menten kinetics.

Since the past few years, much attention was focused on inhibition mathematical models are applied to describe the effect of toxicants on microorganisms [1]. Also, the efficacy of a biofilm has been compared to suspended growth systems which are quantified using effectiveness factor. Biofilms are complex microbial ecosystems in which many biological and chemical processes take place simultaneously. In order to evaluate such systems, mathematical models could be very helpful [2].

Recently, Rao et al. [3] had established mathematical and kinetic modeling of biofilm reactor based on ant colony optimization. Mannina et al. [4] had presented

© Springer Nature Singapore Pte Ltd. 2019

G. Hariharan, *Wavelet Solutions for Reaction–Diffusion Problems in Science and Engineering*, Forum for Interdisciplinary Mathematics,
https://doi.org/10.1007/978-981-32-9960-3_11

the modeling and dynamic simulation of hybrid moving bed biofilm reactors. Wang and Zhang [5] reviewed the progress made in the mathematical modeling of biofilms. The mathematical models addressed in this review article have already been used to explain many complicated phenomena in the biofilm dynamics. Many physical phenomena on the engineering and science fields are frequently modeled by nonlinear differential equations [7]. Such equations are often difficult or impossible to solve analytically.

Based on the successful applications of Chebyshev wavelet spectral method in numerical solution of differential equations, together with the properties of Chebyshev wavelet function, we believe that they can be applicable in solving biofilm-based enzymatic reaction equations. In recent years, wavelet analysis or wavelet method has found their way into many different fields in science, engineering, and medicine. It possesses many useful properties, such as compact support, orthogonality, dyadic, orthonormality, and multiresolution analysis (MRA). Recently, wavelets have been applied extensively for signal processing in communications and physics research and have proved to be a wonderful mathematical tool. After discretizing the differential equations in a conventional way like the finite difference approximation, wavelets can be used for algebraic manipulations in the system of equations obtained which lead to better condition number of the resulting system [6–8]. Also, this technique allows the creation of fast algorithms when compared to the algorithm which is used ordinarily.

There is a growing interest in using various wavelets to study problems, of greater computational complexity. Among the wavelet transform families, the Haar, Legendre, and Chebyshev wavelets deserve much attention. The basic idea of Chebyshev wavelet method (CWM) is to convert the differential equations into a system of algebraic equations by the operational matrices of integral or derivative [9–11]. The main goal is to show how wavelets and multiresolution analysis can be applied for improving the method in terms of easy implementability and achieving the rapidity of its convergence. Wavelets, as very well-localized functions, are considerably useful for solving differential equations and provide accurate solutions. Also, the wavelet technique allows the creation of very fast algorithms when compared with the algorithms ordinarily used [7–9]. Recently, Hariharan and Kannan [10] reviewed the wavelet transforms methods for solving a few reaction–diffusion equations arising in science and engineering. Hariharan and his co-workers [11–15] have introduced the wavelet methods for solving a few reaction–diffusion equations. Several wavelet-based spectral algorithms have been developed to solve the nonlinear differential equations and applications, for instance, the reader is advised to see [16–21] and references therein. In this paper, an efficient wavelet-based analytical expression is obtained for the steady-state concentration of substance and flux into the biofilm using shifted second kind Chebyshev wavelets method (S2KCWM).

This chapter is summarized as follows. In Sect. 11.2, the mathematical formulation of the problem is discussed. In Sect. 11.2.1, the shifted second kind Chebyshev wavelets with some of their properties are presented. The method of solution of the biofilm substrate concentration model is presented in Sect. 11.3. Concluding remarks are given in Sect. 11.4.

11.2 Mathematical Formulation of the Problem [1]

Based on the theory of enzymatic reactions on the particle surface of biofilm, it is assumed that the substrate consumption with the Michaelis–Menten kinetics model formulated as [1]

$$D_f \frac{d^2 S_f}{dz^2} = q \frac{S_f}{K + S_f} X_j \tag{11.1}$$

The boundary conditions are

$$z = 0, \frac{dS_f}{dz} = 0$$
$$z = L_f, S_f = S_1 \tag{11.2}$$

The biomass balance is described (See Ref. [1])

$$Y_q \frac{S_f}{K + S_f} X_f = b X_f^2 \tag{11.3}$$

From Eq. (11.3), the concentration of active biomass can be expressed through the substrate concentration. Now, the Eq. (1.1) can be written in the form

$$D_f \frac{d^2 S_f}{dz^2} = \frac{q^2 Y}{b} \left[\frac{S_f}{K + S_f} \right]^2 \tag{11.4}$$

where S_f is the substrate concentration in the biofilm, K is the Michaelis–Menten constant, z is the coordinate, L_f is the biofilm thickness, D_f is the diffusion coefficient within the biofilm, b is the microbial death constant, q is the substrate consumption rate constant, S_1 is the substrate concentration outside the biofilm, and Y is the biomass yield per unit amount of substrate consumed, respectively. The nonlinear ODE is made dimensionless by defining the following parameters:

$$S = \frac{S_f}{K}, x = \frac{z}{L_f}, \delta = \frac{Y q^2 L_f^2}{b K D_f}, S_L = \frac{S_1}{K}$$

Equation (11.4) reduces to the following dimensionless form:

$$\frac{d^2S}{dx^2} = \delta\left(\frac{S^2}{S^2 + 2S + 1}\right) \tag{11.5}$$

$$\left.\begin{array}{l} x = 0, \frac{ds}{dx} = 0 \\ x = 1, S = S_L = 5 \\ x = 1, S = S_L = 0.5 \end{array}\right\} \tag{11.6}$$

The dimensionless concentration flux into the biofilm is given

$$\varphi(x) = \frac{1}{\sqrt{\delta}} \frac{dS}{dx}\bigg|_{x=1} \tag{11.7}$$

11.2.1 Solution of the Boundary Problem by Shifted Second Kind Chebyshev Wavelets [16]

It is well known that the second kind Chebyshev polynomials are defined on $[-1, 1]$ by

$$U_n(x) = \frac{\sin(n+1)\theta}{\sin\theta}, x = \cos\theta. \tag{11.8}$$

These polynomials are orthogonal on $[-1, 1]$, i.e.,

$$\int_{-1}^{1} \sqrt{1 - x^2}\, U_m(x)U_n(x)\, dx = \begin{cases} 0, m \neq n \\ \frac{\pi}{2} \end{cases} m = n \tag{11.9}$$

The following properties of second kind Chebyshev polynomials are of fundamental importance in the sequel. They are eigenfunctions of the following singular Sturm–Liouville equation.

$$(1 - x^2)D^2\phi_k(x) - 3xD\phi_k(x) + k(k+2)\phi_k(x) = 0, \tag{11.10}$$

where $D \equiv \frac{d}{dx}$ and may be generated by using the recurrence relation

$$U_{k+1}(x) = 2xU_k(x) - U_{k-1}(x), k = 1, 2, 3, \ldots \tag{11.11}$$

Starting from $U_0(x) = 1$ to $U_1(x) = 2x$, or from Rodrigues formula

$$U_n(x) = \frac{(-2)^n(n+1)!}{(2n+1)!\sqrt{(1 - x^2)}}D^n\left[(1 - x^2)^{n+\frac{1}{2}}\right] \tag{11.12}$$

11.2.2   Shifted Second Kind Chebyshev Polynomials (S2KCP)

The shifted second kind Chebyshev polynomials are defined on $[0, 1]$ by $U_n^*(x) = U_n(2x - 1)$. All results of second kind Chebyshev polynomials can be easily transformed to give the corresponding results for their shifted forms. The orthogonality relation with respect to the weight function $\sqrt{x - x^2}$ is given by

$$\int_0^1 \sqrt{x - x^2}\, U_n^*(x) U_m^*(x)\, dx = \begin{cases} 0, & m \neq n \\ \frac{\pi}{8}, & m = n. \end{cases} \tag{11.13}$$

11.2.3   Shifted Second Kind Chebyshev Operational Matrix of Derivatives [16]

Wavelets constitute a family of functions constructed from dilation to translation of a single function called the mother wavelet. When the dilation parameter 'a' and the translation parameter 'b' vary continuously, then we have the following family of continuous wavelets:

$$\psi_{a,b}(t) = |a|^{-1/2} \psi\left(\frac{t - b}{a}\right) \quad a, b \in R, \quad a \neq 0 \tag{11.14}$$

Second kind Chebyshev wavelets; $\psi_{n,m}(t) = \psi(k, n, m, t)$ where k, n can assume any positive integer, m is the order of second kind Chebyshev polynomials, and t is the normalized time. They are defined on the interval $[0, 1]$ by

$$\psi_{n,m}(t) = \begin{cases} \frac{2^{\frac{k+3}{2}}}{\sqrt{\pi}} U_m^*(2^k t - n), & t \in \left[\frac{n}{2^k}, \frac{n+1}{2^k}\right], \\ 0 & \text{otherwise} \end{cases} \tag{11.15}$$

$m = 0, 1, \ldots M,\ n = 0, 1, \ldots 2^k-1$. A function $f(t)$ defined over $[0, 1]$ may be expanded in terms second kind Chebyshev wavelets as

$$f(t) = \sum_{n=0}^{\infty} \sum_{m=0}^{\infty} c_{nm} \psi_{nm}(t), \tag{11.16}$$

where

$$c_{nm} = (f(t), \psi_{nm}(t))_w = \int_0^1 \sqrt{t - t^2} f(t) \psi_{nm}(t) \, dt. \qquad (11.17)$$

If the infinite series is truncated, then it can be written as

$$f(t) = \sum_{n=0}^{\infty} \sum_{m=0}^{\infty} c_{nm} \psi_{nm}(t). = C^{\mathrm{T}} \psi(t), \qquad (11.18)$$

where C and $\psi(t)$ are $2^k(M + 1) \times 1$ defined by

$$\left.\begin{array}{l} C = \left[c_{0,0}, c_{0,1}, \ldots c_{0,M}, \ldots, c_{2^k-1,M}, \ldots c_{2^k-1,1}, \ldots, c_{2^k-1,M}\right]^{\mathrm{T}} \\ \psi(t) = \left[\psi_{0,0}, \psi_{0,1}, \ldots, \psi_{0,M}, \ldots \psi_{2^k-1,M}, \ldots, \psi_{2^k-1,1}, \ldots, \psi_{2^k-1,M}\right]^{\mathrm{T}} \end{array}\right\} \quad (11.19)$$

11.3 Method of Solution

In this section, the S2KCWM is introduced for solving the model problem. Using the proposed algorithm described in Sect. 11.2 for $M = 2$, $k = 0$ and obtain an approximate solution of $S(x)$.

Case (i) Consider the nonlinear equation Eq. (11.5)

$$\frac{d^2 S}{dx^2} = \delta \left(\frac{S}{S+1}\right)^2$$

with the conditions $\left.\begin{array}{l} x = 0, \frac{dS}{dx} = 0 \\ x = 1, S = S_{\mathrm{L}} = 5 \end{array}\right\}$ Equation (11.5) can be written as

$$\left(1 + C^{\mathrm{T}} \psi(x)\right)^2 C^{\mathrm{T}} D^2 \psi(x) = \delta \left(C^{\mathrm{T}} \psi(x)\right)^2 \qquad (11.20)$$

Using the Chebyshev wavelet operational matrices,

$$D = \begin{bmatrix} 0 & 0 & 0 \\ 4 & 0 & 0 \\ 0 & 8 & 0 \end{bmatrix}, \quad D^2 = \begin{bmatrix} 0 & 0 & 0 \\ 0 & 0 & 0 \\ 32 & 0 & 0 \end{bmatrix}$$

Moreover $\psi(x)$ can be obtained as $\psi(x) = \sqrt{\frac{2}{\pi}}\begin{bmatrix} 2 \\ 8x - 4 \\ 32x^2 - 32x + 6 \end{bmatrix}$ If we set

$C = \begin{pmatrix} c_{0,0} & c_{0,1} & c_{0,2} \end{pmatrix}^{\mathrm{T}} = \sqrt{\frac{\pi}{2}}\begin{pmatrix} c_0 & c_1 & c_2 \end{pmatrix}^{\mathrm{T}}$, then Eq. (11.5) takes the form

$$\left(1 + 2c_0 + c_1(8x - 4) + c_2(32x^2 - 32x + 6)\right)^2$$
$$= \delta\left(2c_0 + c_1(8x - 4) + c_2(32x^2 - 32x + 6)\right)^2 \tag{11.21}$$

We only need to satisfy the equation at the first root $U_3^*(x)$, i.e., $x = \frac{2-\sqrt{2}}{4}$, and we gain

$$\left(2c_o - 2\sqrt{2}c_1 + 2c_2 + 1\right)^2 64c_2 = \delta\left(2c_o - 2\sqrt{2}c_1 + 2c_2\right)^2 \tag{11.22}$$

Using the initial conditions

$$c_1 - 4c_2 = 0 \tag{11.23}$$

$$2c_0 + 4c_1 + 6c_2 = 5 \tag{11.24}$$

$$2c_0 + 4c_1 + 6c_2 = 0.5 \tag{11.25}$$

Using the Adomian decomposition method (ADM), the solution of Eq. (11.5) is given by [1]

$$S(x) = S_L + \frac{\delta}{2}\left[\frac{S_L}{1+S_L}\right]^2 (x^2 - 1) + \frac{\delta^2 S_L^3}{(1+S_L)^5}\left[\frac{x^4}{12} - \frac{x^2}{2} + \frac{5}{12}\right] + \frac{\delta^2 S_L^2}{4(1+S_L)^6}\left[\frac{x^6}{30} - \frac{x^4}{6} + \frac{x^2}{2}\right]$$
$$- \frac{\delta^2 S_L^5}{(1+S_L)^7}\left[\frac{x^6}{30} - \frac{x^4}{6} + \frac{x^2}{2}\right] + \frac{2\delta^2 S_L^2}{(1+S_L)^7}\left[\frac{x^6}{360} - \frac{x^4}{24} + \frac{5x^2}{24}\right] + \frac{3\delta^2 S_L^6}{4(1+S_L)^8}\left[\frac{x^6}{30} - \frac{x^4}{6} + \frac{x^2}{2}\right]$$
$$- \frac{2\delta^2 S_L^5}{(1+S_L)^8}\left[\frac{x^6}{360} - \frac{x^4}{24} + \frac{5x^2}{24}\right] + \frac{2\delta^2 S_L^5}{(1+S_L)^8}(-155 + 66S_L)$$

$$\tag{11.26}$$

The solution of concentration flux into the biofilm is obtained as

$$\psi = \frac{1}{\sqrt{\delta}}\left[\begin{matrix} \delta\left[\frac{S_L}{1+S_L}\right]^2 - \frac{2}{3}\frac{\delta^2 S_L^3}{(1+S_L)^5} + \frac{2}{15}\frac{\delta^2 S_L^2}{(1+S_L)^6} - \frac{8}{15}\frac{\delta^2 S_L^5}{(1+S_L)^7} \\ + \frac{8}{15}\frac{\delta^2 S_L^2}{(1+S_L)^7} + \frac{2}{5}\frac{\delta^2 S_L^6}{(1+S_L)^8} - \frac{8}{15} + \frac{2}{5}\frac{\delta^2 S_L^5}{(1+S_L)^8} \end{matrix}\right] \tag{11.27}$$

Tables 11.1 and 11.2 show the accuracy of the proposed method.

Table 11.1 Comparison of normalized steady-state concentration $S(x)$ with S2KCWM results for various values of x and δ (Eq. 11.5)

Concentration of $S(x)$ when $S_L = 5$

x	$S(x)$ (when $\delta = 1$)			$S(x)$ (when $\delta = 5$)			$S(x)$ (when $\delta = 10$)		
	Simulation	ADM	S2KCWM	Simulation	ADM	S2KCWM	Simulation	ADM	S2KCWM
0.0	4.6500	4.6600	4.6611	3.4500	3.4520	3.4514	2.4000	2.3330	2.3377
0.2	4.67 00	4.6730	4.6731	3.5140	3.5140	3.5130	2.4450	2.4340	2.4443
0.4	4.7120	4.7150	4.7143	3.6990	3.6990	3.6991	2.6500	2.7400	2.7643
0.6	4.7800	4.7800	4.7820	3.9920	3.9920	4.0088	3.2600	3.2660	3.2977
0.8	4.8760	4.8760	4.8770	4.5300	4.5300	4.4425	4.0020	4.0310	4.0445
1.0	5.0000	5.0000	5.0000	5.0000	5.0000	5.0000	5.0000	5.0000	5.0000

Table 11.2 Comparison of normalized steady-state concentration $S(x)$ with S2KCWM results for various values of x and δ

Concentration of $S(x)$ when $S_L = 0.5$

x	$S(x)$ (when $\delta = 0.5$)			$S(x)$ (when $\delta = 1$)			$S(x)$ (when $\delta = 3$)		
	Simulation	ADM	S2KCWM	Simulation	ADM	S2KCWM	Simulation	ADM	S2KCWM
0.0	0.4738	0.4738	0.4738	0.4506	0.4506	0.4504	0.3785	0.3785	0.3780
0.2	0.4748	0.4748	0.4748	0.4520	0.4523	0.4523	0.3831	0.3831	0.3828
0.4	0.4780	0.4780	0.4779	0.4579	0.4581	0.4583	0.3969	0.3971	0.3975
0.6	0.4831	0.4834	0.4832	0.4670	0.4678	0.4682	0.4210	0.4211	0.4217
0.8	0.4900	0.4900	0.4905	0.4815	0.4818	0.4821	0.4557	0.4559	0.4560
1.0	0.5000	0.5000	0.49999	0.5000	0.5000	0.5000	0.5000	0.5000	0.5000

Case (ii) Consider the nonlinear equation Eq. (11.5)

$$\frac{d^2 S}{dx^2} = \delta \left(\frac{S}{S+1} \right)^2$$

with the conditions $\left. \begin{array}{l} x = 0, \frac{ds}{dx} = 0 \\ x = 1, S = S_{\mathrm{L}} = 0.5 \end{array} \right\}$ Equation (11.5) can be written as

$$\left(1 + C^{\mathrm{T}} \psi(x) \right)^2 C^{\mathrm{T}} D^2 \psi(x) = \delta \left(C^{\mathrm{T}} \psi(x) \right)^2$$

An approximate analytical expression of concentration S is given in Eq. (11.26) (see Ref. [1]). The concentration S is plotted in figures for various values of δ and S_{L} (see Figs. 11.1, 11.2, 11.3, 11.4, 11.5, and 11.6). From these figures, it is evident that the value of concentration gradually increases as the dimensionless biofilm thickness δ decreases. From these figures, it is observed that the value of the concentration increases when S_{L} increases. When $\delta \leq 1$, the concentration is uniform, and the uniform value depends upon S_{L}. It is clear that dimensionless substrate concentration outside the biofilm S_{L} increases when the value of the dimensionless concentration S (x) increases. Our numerical results can be compared with Muthukaruppan et al. [1] results. Good agreement with the ADM and simulation results is observed. For larger M, we can get the results closer to the real values.

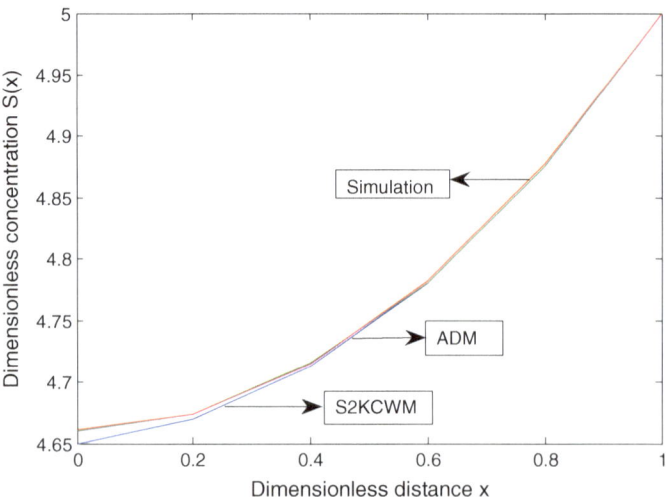

Fig. 11.1 Normalized dimensionless concentration $S(x)$ with dimensionless distance x when the biofilm thickness $\delta = 1$

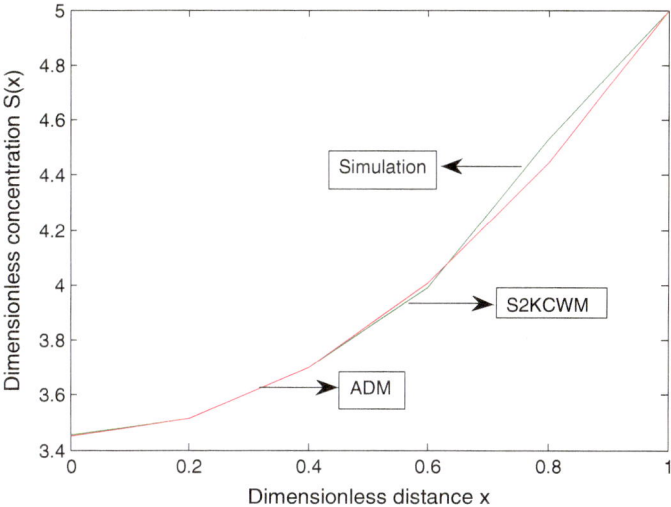

Fig. 11.2 Normalized dimensionless concentration $S(x)$ with dimensionless distance x when the biofilm thickness $\delta = 5$

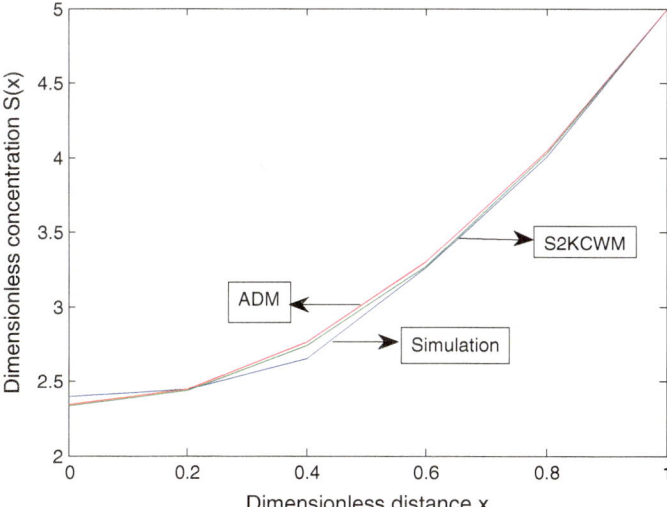

Fig. 11.3 Normalized dimensionless concentration $S(x)$ with dimensionless distance x when the biofilm thickness $\delta = 10$

Figures 11.1, 11.2, 11.3, 11.4, 11.5, and 11.6 show the numerical results for various values of x and δ by the S2KCWM for the examples. It has been well demonstrated that in applying the nice properties of Chebyshev wavelets, the

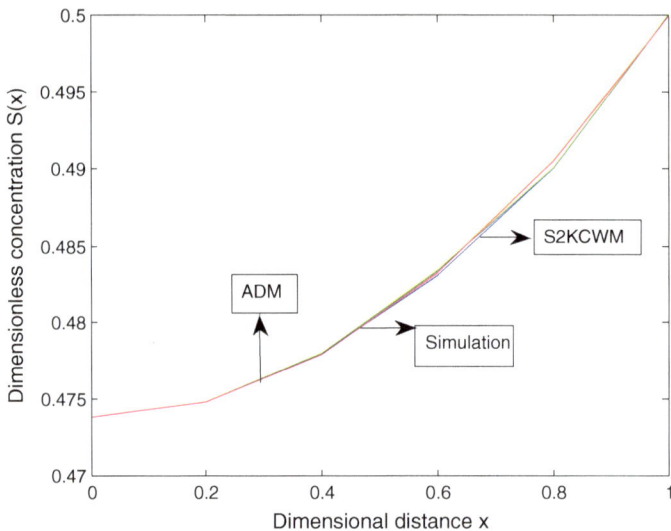

Fig. 11.4 Normalized dimensionless concentration $S(x)$ with dimensionless distance x when the biofilm thickness $\delta = 0.5$

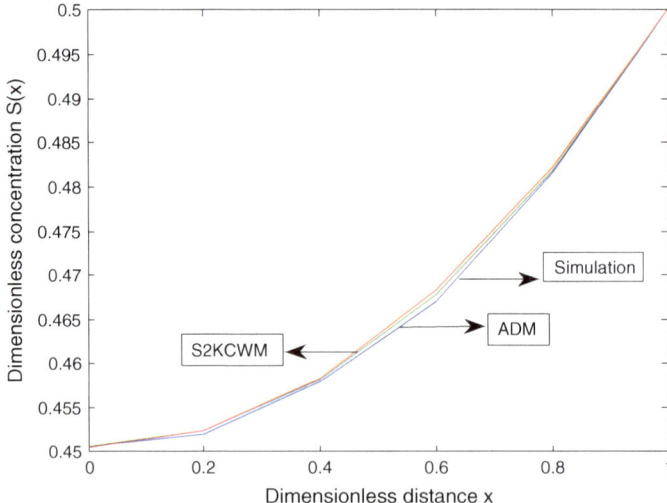

Fig. 11.5 Normalized dimensionless concentration $S(x)$ with dimensionless distance x when the biofilm thickness $\delta = 1$

differential equation can be solved conveniently by using S2KCWM systematically. The method with far less degrees of freedom and with smaller CPU time provides better solutions than classical ones. The accuracy and effectiveness of the method

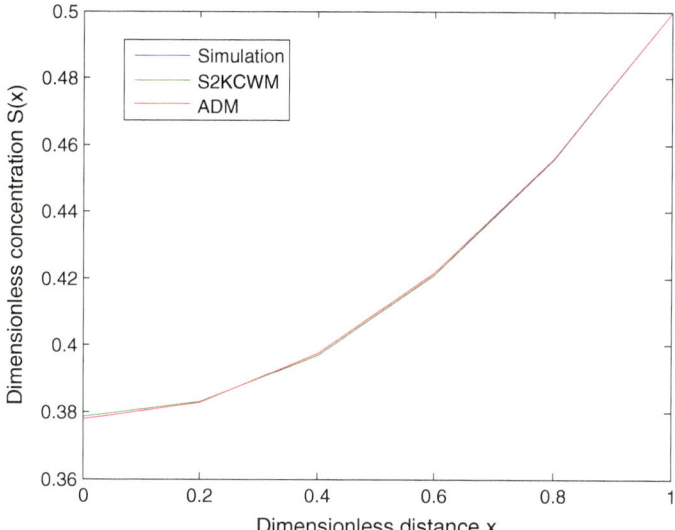

Fig. 11.6 Normalized dimensionless concentration $S(x)$ with dimensionless distance x when the biofilm thickness $\delta = 3$

are analyzed; the results obtained are compared with the ADM and simulation results. The power of the manageable method is confirmed.

All the numerical experiments presented in this section were computed in double precision with some MATLAB codes on a personal computer System Vostro 1400 Processor x86 Family 6 Model 15 Stepping 13 Genuine Intel $\sim$1596 MHz.

11.4 Concluding Remarks

A nonlinear steady-state reaction–diffusion (RDEs) of the biofilm model has been solved by the Chebyshev wavelets method (CWM) analytically. The efficiency and accuracy of the proposed method have been presented by using the illustrative examples. In fact, the proposed wavelet method provides a direct scheme for obtaining the approximation of the solution. The proposed wavelet-based analytical method is used for analyzing biofilm for a square law of microbial death rate. Theoretical model is discussed for an investigation of the dynamic behavior of substrate consumption by a biofilm. The wavelet based analytical expressions for the steady state substrate concentration for all values of biochemical parameters δ and S_L. Also steady-state flux response is also discussed. Good agreement with the ADM and simulation results is observed. This predicted model is very useful to further develop the model that involves the balance of production of active biomass and biofilm erosion. The proposed wavelet method is capable for solving nonlinear

BVPs arising in chemical sciences. It may be concluded that shifted second kind CWT is very powerful and finding the analytical solution as well as the numerical solution for a wide class of linear and nonlinear differential equations.

References

1. S. Muthukaruppan, A. Eswari, Lakshmanan Rajendran, Mathematical modeling of a biofilm: the adomian decomposition method. Nat. Sci. **5**(4), 456–462 (2013)
2. S. Usha, S. Anitha, L. Rajendran, Approximate analytical solution of non-linear reaction diffusion equation in fluidized bed biofilm reactor. Nat. Sci. **12**(4), 983–991 (2012)
3. K.R. Rao, T. Srinivasan, C. Venkateswarlu, Mathematical and kinetic modeling of biofilm reactor based on ant colony optimization. Proc. Biochem. **45**(6), 961–972 (2010)
4. G. Mannin, D. Di Trapani, G. Viviani, H. Ødegaard, Modelling and dynamic simulation of hybrid moving bed biofilm reactors: Model concepts and application to a pilot plant. Biochem. Eng. J. **56**(1–2), 23–36 (2011)
5. Q. Wang, T.Y. Zhang, Review of mathematical models for biofilms. Sol. State Commun. **150**(21), 1009–1022 (2010)
6. D. Mary Celin Sharmila, T. Praveen, L. Rajendran, Mathematical modeling and analysis of nonlinear enzyme catalyzed reaction processes, vol. 2013, Article ID 931091, p. 7
7. P. Pirabaharan, R.D. Chandrakumar, G. Hariharan, An efficient wavelet based approximation method for estimating the concentration of species and effectiveness factors in porous catalysts. MATCH **73**(3), 705–727 (2015)
8. M. Sivasankari, L. Rajendran, Analytical expressions of steady-state concentrations of species in potentiometric and amperometric biosensor. Nat. Sci. **4**(12), 1029–1041 (2012)
9. A. KazemiNasab, A. Kilieman, E. Babolian, Z.Pashazadeh Atabakan, Wavelet analysis method for solving linear and nonlinear singular boundary value problems. Appl. Math. Model. **37**, 5876–5886 (2013)
10. Farikhin, I. Mohd, Orthogonal functions based on Chebyshev polynomials. Matematika **27**(1), 97–107 (2011)
11. S.G. Hosseini, A new operational matrix of derivative for Chebyshev wavelets and its applications in solving ordinary differential equations. Appl. Math. Sci. **5**(51), 2537–2548 (2011)
12. G. Hariharan, K. Kannan, Haar wavelet method for solving nonlinear parabolic equations. J. Math. Chem. **48**, 1044–1061 (2010)
13. G. Hariharan, K. Kannan, K.R. Sharma, Haar wavelet in estimating the depth profile of soil temperature. Appl. Math. Comput. **210**, 119–225 (2009)
14. G. Hariharan, K. Kannan, Haar wavelet method for solving Fisher's equation. Appl. Math. Comput. **211**, 284–292 (2009)
15. G. Hariharan, K. Kannan, A comparative study of a haar wavelet method and a restrictive Taylor's series method for solving convection-diffusion equations. Int. J. Comput. Methods Eng. Sci. Mech. **11**(4), 173–184 (2010)
16. G. Hariharan, K. Kannan, Review of wavelet methods for the solution of reaction diffusion problems in science and engineering. Appl. Math. Modell. (Press, 2013)
17. E.H. Doha, A.H. Bhrawy, S.S. Ezz Eldien, A Chebyshev spectral method based on operational matrix for initial and boundary value problems of fractional order. Comput. Math Appl. **62**, 2364–2373 (2011)
18. E.H. Doha, A.H. Bhrawy, S.S. Ezz Eldien, Efficient Chebyshev spectral methods for solving multi-term fractional orders differential equations. Appl. Math. Modell. **35**(2011), 5662–5672 (2011)

19. E.H. Doha, A.H. Bhrawy, M.A. Saker, Integrals of Bernstein polynomials: an application for the solution of high even-order differential equations. Appl. Math. Lett. **24**, 559–565 (2011)
20. A. Saadatmandi, M. Dehghan, A Legendre collocation method for fractional integro-differential equations. J. Vibr. Contr. **17**, 2050–2058 (2011)
21. S. Esmaeili, M. Shamsi, A pseudo –spectral scheme for the approximate solution of a family of fractional differential equations. Commun. Nonlinear Sci. Numer. Simulat. **16**, 3646–3654 (2011)